Automatic Control and Routing of Marine Vessels

Automatic Control and Routing of Marine Vessels

Editors

Evgeny Veremey
Margarita Sotnikova

MDPI • Basel • Beijing • Wuhan • Barcelona • Belgrade • Manchester • Tokyo • Cluj • Tianjin

MDPI

Editors
Evgeny Veremey
Saint Petersburg State University
Russia

Margarita Sotnikova
Saint Petersburg State University
Russia

Editorial Office
MDPI
St. Alban-Anlage 66
4052 Basel, Switzerland

This is a reprint of articles from the Special Issue published online in the open access journal *Journal of Marine Science and Engineering* (ISSN 2077-1312) (available at: https://www.mdpi.com/journal/jmse/special_issues/Evgeny_automatic_control_routing_marine_vessels).

For citation purposes, cite each article independently as indicated on the article page online and as indicated below:

LastName, A.A.; LastName, B.B.; LastName, C.C. Article Title. *Journal Name* **Year**, *Volume Number*, Page Range.

ISBN 978-3-0365-5919-3 (Hbk)
ISBN 978-3-0365-5920-9 (PDF)

Contents

Journal of
Marine Science and Engineering

MDPI

Editorial

Automatic Control and Routing of Marine Vessels

Margarita V. Sotnikova

Department of Computer Applications and Systems, Saint Petersburg State University, 198504 Saint Petersburg, Russia; m.sotnikova@spbu.ru

1. Introduction

Due to the intensive development of the global economy, many problems are constantly emerging connected with the safety of ships' motion in the context of increasing marine traffic. These problems seem to be especially significant for the further development of marine transportation services, with the need to considerably increase their efficiency and reliability. One of the commonly used approaches to ensuring safety and efficiency is the wide implementation of various automated systems for guidance and control, including popular systems such as marine autopilots, dynamic positioning systems, speed control systems, automatic routing installations, etc.

This Special Issue is focused on various problems related to the analysis, design, modelling, and operation of the aforementioned systems. The Issue collected ten papers that cover such areas as identification using neural networks, optimal weather routing, sliding mode control, tracking control, logistics and cooperation between seaports and carriers, control of multi-joint autonomous underwater vehicles, path-following control, and collision avoidance systems.

A brief description of each paper is given in the following section.

2. Papers Details

Xu et al. [1] proposed using a physics-informed neural network (PINN) to identify the dynamic models of the unmanned surface vehicle (USV). PINN has an advantage of combining the data-driven machine learning and physical models. It is shown how PINN can be adopted to embed the dynamic models of an USV into the loss function. The special tests were carried out in the Qing Huai river to obtain the empirical data. The PINN method was implemented and compared with a tradition neural network. The results of the investigation indicate that the PINN has a better performance in predicting the USV dynamics under a small number of training samples.

González-Prieto et al. [2] investigated the course keeping control problem for a USV in the presence of unknown disturbances and system uncertainties. The authors developed the adaptive integral non-linear controller based on the sliding mode surface with adaptive gains. The mathematical background for the proposed control design approach is provided. Various numerical simulations have been carried out with fixed and time-varying references and different external disturbances. The obtained results show that the developed controller allow to achieve the desired performance of the closed-loop system in course keeping problem.

Lezhnina et al. [3] analyzed the effectiveness of cooperation between ports and carriers in the logistics chain in order to increase profits and reduce costs. Special emphasis is placed on the role of peripheral ports, which give ports an opportunity to increase the level of loading and unloading services, and allows sea routes to receive a stable income from the port's operation. The presented study uses a game-theoretic approach and describes a multi-port game scenario. A mathematical model is built and the algorithm for optimal route search is proposed. Numerical experiments are shown that indicate the effectiveness

Citation: Sotnikova, M.V. Automatic Control and Routing of Marine Vessels. *J. Mar. Sci. Eng.* **2022**, *10*, 618. https://doi.org/10.3390/jmse10050618

Received: 25 April 2022
Accepted: 28 April 2022
Published: 1 May 2022

of cooperation in reducing the effects of disruptions as well as reducing overall logistics costs.

Zhao et al. [4] presented a multicriteria ship route planning method based on improved particle swarm optimization–genetic algorithm. This algorithm aims to optimize the meteorological risk, fuel consumption, and navigation time of a ship. The output of the algorithm allows to obtain the minimum-navigation-time route, the minimum-fuel-consumption route, the minimum-navigation-risk route, and the recommended route. The simulation experiments for a container ship are provided. These experiments show the feasibility and effectiveness of the proposed approach.

Yu et al. [5] discussed the problem of modeling and control of multi-joint autonomous underwater vehicles (MJ-AUV). The considered MJ-AUV is a multi-input and multi-output system, where the variation of two joint angles is the input and the pitch and yaw angles of the body is the output. The 3D motion model of the MJ-AUV was established using Newton's second law and the principle of moment balance. In order to design feedback control, the system was decoupled into pitch and yaw subsystems. The linear state observer and LQR design approach was used to obtain optimal control law for each subsystem. The numerical simulation was carried out to verify the obtained mathematical model and to test the effectiveness of the control algorithm.

Volkova et al. [6] analyzed the possibility of using neural networks to predict the coordinates of the vessel during river navigation. This problem is of practical importance, in particular, for autonomously moving vessels. The proposed predicting system can serve as a source of additional information for timely decision-making in case of AIS signal distortion. The real experiment was conducted to collect data, which was used to train neural networks with different activation functions and different structures. A number of simulation experiments were carried out and discussed to assess the reliability of the obtained results for various trajectories of the vessel.

Zhang et al. [7] proposed a new path-following control law. The main innovation is a developed hyperbolic guidance law, which is used to control a marine ship on a straight-line path. The authors have shown that this approach allow to improve the performance characteristics of the closed-loop system in comparison with existing control algorithms. Also, a problem of curved path-following is considered, where the curve is formed as a transition between two adjacent straight-line paths. The modification of the reverse stepping method is proposed to make the system globally asymptotically stable. The simulation experiments are carried out for three different types of ships and the control effect is evaluated and discussed.

Songtao et al. [8] studied the use of fin stabilizers to reduce the rolling and heeling during ship turning. The authors have established the nonlinear mathematical model of a vessel, taking into account forces and moments produced by fin stabilizers, rudders, propellers, and waves. The resulting nonlinear control model has uncertainty due to inaccuracy of parameters and external disturbances. For this reason, the L2-gain based adaptive robust method is proposed to control the fin stabilizers. The controller design process and the proof of the closed-loop system stability are presented in details. The simulation experiments are carried out, where the comparison with the well-tuned PID controller is given and discussed.

Veremey [9] investigated the design of stabilizing feedback control laws for marine vessels moving along initially given trajectories. Unlike the traditional methods, it is proposed to use an optimization approach and implement the optimal damping (OD) concept, previously developed by V.I. Zubov. Essential attention is paid to the practical adaptation of the optimal damping methods for marine control systems. A new method for tracking controllers' design, which ensures the desirable reference motion of the vessel along the forward speed and heading angle is proposed. A new methodology for selecting the functional to be damped is discussed. The choice of this functional influences the properties of the closed-loop system, such as asymptotic stability and the quality of control processes. The important feature is that the developed OD based approach can be implemented in real-time regime of a ship's motion. The practical applicability and effectiveness of the approach is illustrated by a numerical example of tracking control design.

Guan et al. [10] discussed issues related to the problem of preventing collisions of ships. The main purpose of the research is to develop a decision-making system to assist sailors and reduce the number of maritime accidents. The paper proposes a ship domain model, which is based on the fuzzy logics and can combine many factors affecting ship collision risk. These factors are determined by ship own parameters and external environmental factors. The composition fuzzy inference is used to determine the range of the ship domain and, as a consequence, to calculate the collision risk. The simulation experiments were conducted to demonstrate the feasibility and effectiveness of the proposed method.

Funding: This research was funded by RFBR, project number 20-07-00531.

Data Availability Statement: All relevant data and links to that can be found in the presented papers at https://www.mdpi.com/journal/jmse/special_issues/Evgeny_automatic_control_routing_marine_vessels (accessed on 24 April 2022).

Acknowledgments: I wish to express my sincere gratitude to all the authors and the reviewers. The work on this Special Issue was one of the last works of Evgeny Veremey. He was doing this work, but, with great regret, did not have the opportunity to complete this and many other projects.

Conflicts of Interest: The authors declare no conflict of interest.

References

1. Xu, P.-F.; Han, C.-B.; Cheng, H.-X.; Cheng, C.; Ge, T. A Physics-Informed Neural Network for the Prediction of Unmanned Surface Vehicle Dynamics. *J. Mar. Sci. Eng.* **2022**, *10*, 148. [CrossRef]
2. González-Prieto, J.A.; Pérez-Collazo, C.; Singh, Y. Adaptive Integral Sliding Mode Based Course Keeping Control of Unmanned Surface Vehicle. *J. Mar. Sci. Eng.* **2022**, *10*, 68. [CrossRef]
3. Lezhnina, E.A.; Balykina, Y.E. Cooperation between Sea Ports and Carriers in the Logistics Chain. *J. Mar. Sci. Eng.* **2021**, *9*, 774. [CrossRef]
4. Zhao, W.; Wang, Y.; Zhang, Z.; Wang, H. Multicriteria Ship Route Planning Method Based on Improved Particle Swarm Optimization–Genetic Algorithm. *J. Mar. Sci. Eng.* **2021**, *9*, 357. [CrossRef]
5. Yu, L.; Meng, Q.; Zhang, H. 3-Dimensional Modeling and Attitude Control of Multi-Joint Autonomous Underwater Vehicles. *J. Mar. Sci. Eng.* **2021**, *9*, 307. [CrossRef]
6. Volkova, T.A.; Balykina, Y.E.; Bespalov, A. Predicting Ship Trajectory Based on Neural Networks Using AIS Data. *J. Mar. Sci. Eng.* **2021**, *9*, 254. [CrossRef]
7. Zhang, Z.; Zhao, Y.; Zhao, G.; Wang, H.; Zhao, Y. Path-Following Control Method for Surface Ships Based on a New Guidance Algorithm. *J. Mar. Sci. Eng.* **2021**, *9*, 166. [CrossRef]
8. Songtao, Z.; Peng, Z. L2-Gain Based Adaptive Robust Heel/Roll Reduction Control Using Fin Stabilizer during Ship Turns. *J. Mar. Sci. Eng.* **2021**, *9*, 89. [CrossRef]
9. Veremey, E.I. Optimal Damping Concept Implementation for Marine Vessels' Tracking Control. *J. Mar. Sci. Eng.* **2021**, *9*, 45. [CrossRef]
10. Guan, Z.; Wang, Y.; Zhou, Z.; Wang, H. Research on Early Warning of Ship Danger Based on Composition Fuzzy Inference. *J. Mar. Sci. Eng.* **2020**, *8*, 1002. [CrossRef]

Journal of
Marine Science and Engineering

MDPI

Article

A Physics-Informed Neural Network for the Prediction of Unmanned Surface Vehicle Dynamics

Peng-Fei Xu [1,2,*], Chen-Bo Han [2], Hong-Xia Cheng [2], Chen Cheng [2] and Tong Ge [1]

1 School of Naval Architecture, Ocean and Civil Engineering, Shanghai Jiao Tong University, Shanghai 200240, China; tongge@sjtu.edu.cn
2 Institute of Marine Vehicle and Underwater Technology, College of Harbor, Coastal and Offshore Engineering, Hohai University, Nanjing 210098, China; 191303060003@hhu.edu.cn (C.-B.H.); chenghongxia@hhu.edu.cn (H.-X.C.); gordanchan@hhu.edu.cn (C.C.)
* Correspondence: xupengfei@hhu.edu.cn

Abstract: A three-degrees-of-freedom model, including surge, sway and yaw motion, with differential thrusters is proposed to describe unmanned surface vehicle (USV) dynamics in this study. The experiment is carried out in the Qing Huai River and the data obtained from different zigzag trajectories are filtered by a Gaussian filtering method. A physics-informed neural network (PINN) is proposed to identify the dynamic models of the USV. PINNs combine the advantages of data-driven machine learning and physical models. They can also embed the speed and steering models into the loss function, which can significantly retain all types of information. Compared with traditional neural networks, the results show that the PINN has better generalization ability in predicting the surge and sway velocities and rotation speed with only limited training data.

Keywords: unmanned surface vehicle (USV); system identification; traditional neural network; physics-informed neural network; zigzag test

Citation: Xu, P.-F.; Han, C.-B.; Cheng, H.-X.; Cheng, C.; Ge, T. A Physics-Informed Neural Network for the Prediction of Unmanned Surface Vehicle Dynamics. *J. Mar. Sci. Eng.* **2022**, *10*, 148. https://doi.org/10.3390/jmse10020148

Academic Editors: Margarita Sotnikova and Evgeny Veremey

Received: 5 December 2021
Accepted: 10 January 2022
Published: 24 January 2022

1. Introduction

In recent years, the unmanned surface vehicle (USV) has attracted considerable attention. The most important advantages of USVs are that they can be employed in extremely dangerous environments compared with traditional manned vehicles. In addition, USVs play a significant role in both commercial and military fields, such as resource exploration [1], shipping [2], mine countermeasures [3] and reconnaissance [4]. To guarantee that USVs can operate with good performance in these fields, the utilization of robust and effective maneuvering controllers is particularly important.

Numerous studies have focused on the control of USVs. System identification is the most popular method for handling indescribable systems from the given input/output data. Nagumo and Noda (1967) utilized continuous least square estimation based on an error-correction training procedure for system identification [5]. Holzhuter (1989) adopted recursive least square estimation in the identification of ship dynamics [6]. Kallstrom and Astrom (1981) demonstrated recursive maximum likelihood estimation used in the ship steering motion and showed a good prediction [7]. Yoon and Rhee (2003) proposed the extended Kalman filter technique and the modified Bryson–Frazier formulation smoother to predict motion variables, hydrodynamic force, vehicle speed and current direction [8]. Shin et al. (2017) utilized particle swarm optimization (PSO) with an adaptive control algorithm to predict the trajectory of USVs [9]. In addition, Selvam (2005) presented a frequency domain identification system for linear steering equations for ship maneuvering in calm seas [10].

Owing to the wind, current and other random disturbances, the modelling of ship dynamics is a strongly nonlinear problem. Khalid et al. (2020) proposed a non-singular adaptive integral-type finite time tracking control (FTSMC) for nonlinear systems with external disturbances [11]. The obtained results showed that the FTSM control technique guarantees that when the switching surface is reached, tracking errors converge to zero at a fast convergence rate. Vu et al. (2020) utilized the robust station-keeping (SK) control algorithm based on a sliding mode control (SMC) theory, designed to guarantee stability and better performance of a hovering over-actuated autonomous underwater vehicle (HAUV) despite the existence of model uncertainties and ocean current disturbance in the horizontal plane (HP) [12]. Thanh et al. (2020) presented a lumped perturbation observer-based robust control method using an extended multiple sliding surface for a system with matched and unmatched uncertainties [13]. An artificial neural network (ANN) that is capable of solving this nonlinear problem has been presented in recent years. Rajesh and Bhattacharyya (2008) proposed an ANN method to deal with the system identification of a large tanker. The Levenberg–Marquardt algorithm was utilized to train the net, and different numbers of hidden neurons were compared to select the best choice for the construction of the ANN [14]. Oskin et al. (2013) presented a recurrent neural network to identify both the linear and nonlinear behavior of ship dynamics [15]. Pan et al. (2013) utilized an efficient neural network approach to track the motion of autonomous surface vehicles with unknown ship dynamics [16]. However, traditional neural networks have some drawbacks and can be considered as "black box" models that cannot interpret the inherent laws and are not able to guarantee the generalization ability. In addition, a large number of training data are also required to obtain the optimized neural network.

In this study, a physics-informed neural network (PINN) is proposed to predict USV dynamics. Under the condition of a small amount of training data, the model can automatically meet the physical constraints and therefore has better generalization performance while ensuring accuracy and can predict the important physical parameters of the model. PINNs embed the dynamic models of an USV into the loss function rather than the pure data. Furthermore, this is the first time a PINN with a large optimal method (Xavier method, Adam method and Resnet block) has been used to identify USV models, which demonstrates that this method is better than traditional neural networks. The hydrodynamic coefficients can be calculated by limited training data directly. This boosts the application of artificial intelligent (AI) in marine research.

The structure of this study is as follows. Section 2 introduces the basic information of the studied USV and describes its dynamic models. In Section 3, the PINN and optimized method are introduced. The experiments of the USV are carried out and used in Section 4 as training and validation samples for system identifications of USV dynamic models and comparisons between the PINN and traditional neural networks. Section 5 concludes this study.

2. USV Dynamics

2.1. Deepsea Warriors Uboat (DW-Uboat)

The DW-uBoat (see Figure 1) produced by the Institute of Marine Vehicle and Underwater Technology of Hohai University, China, is designed as a streamline body due to anti-resistance. The boat has large storage to carry various instruments and equipment, as well as a large cover on the deck to disassemble and maintain the equipment. The vehicle is installed with two electric propellers that replace the traditional rudder and gimbaled thruster at the stern side of each hull. The steering motions are generated by the rpm differences in the two main thrusters. The length of the DW-uBoat is 1.80 m, the width is 0.70 m and the height is 0.48 m. The maximum speed of the vehicle is 2.5 knots. Moreover, there is no rudder or gimbaled thruster equipped in this vehicle, the steering motion can be generated by the operating RPM (revolution per minute) difference in the two main thrusters, the sway force action on the vehicle is also produced by the difference of the thrusters. Various sensors are equipped in the DW-uBoat as well. Two differential GPSs are installed in the bow and stern, respectively, combined with an electronic compass which can obtain the position, velocity and heading angle information. All of the above information and data are transformed into the computer during the experiments.

Figure 1. Profile of the DW-uBoat.

2.2. Assumption

In order to simply describe the motion of the USV, the three degrees of freedom (surge, sway and yaw) in the horizontal planar motion are considered. The notation proposed by Fossen (2002) was utilized and the description of the USV with its coordinate frame is shown in Figure 2, where u, v and V denote the surge, sway and total velocities of the USV in a body fixed frame, respectively, while x_i and y_i denote the north and east directions of the USV in the inertial frame, respectively. In addition, ψ and β denote the heading angle and course angle of the USV, respectively, while X represents the side slip angle. These simplifications can be concluded as follows:

- The motion of the USV in roll, pitch, and heave directions was neglected.
- The USV had neutral buoyancy and the origin of the body-fixed coordinate was located at the center of mass.
- The dynamic equations of the USV did not include the disturbance forces (waves, wind, and ocean currents).

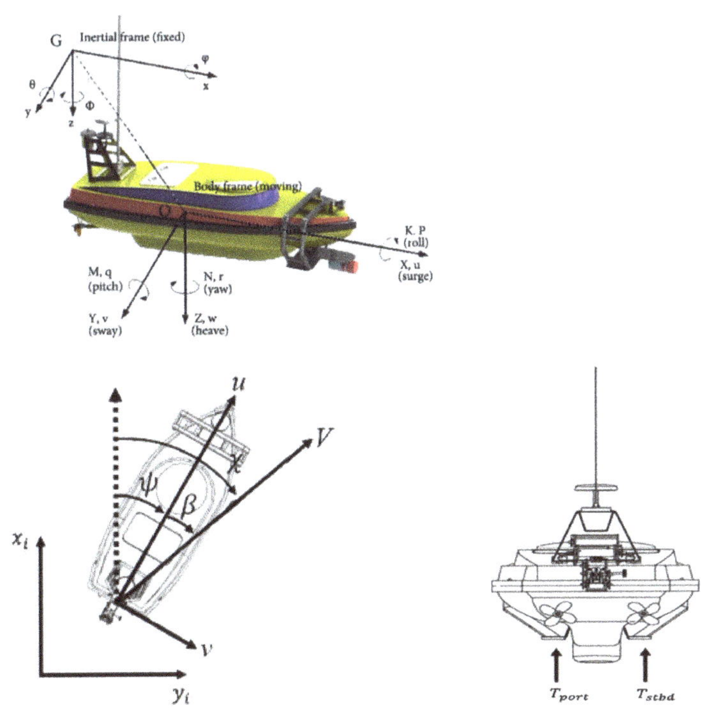

Figure 2. Schematic description of differential thrust of the USV.

2.3. Dynamic Models

The equations of the kinematic model of the USV can be expressed as (Woo et al., 2018):

$$\dot{\boldsymbol{\eta}} = \mathbf{R}(\boldsymbol{\eta}) \cdot \mathbf{v} \tag{1}$$

$$\mathbf{v} = (u, v, r)^T \tag{2}$$

$$\boldsymbol{\eta} = (x_i, y_i, \psi) \tag{3}$$

$$\mathrm{R}(\boldsymbol{\eta}) = \begin{bmatrix} \cos(\psi) & -\sin(\psi) & 0 \\ \sin(\psi) & \cos(\psi) & 0 \\ 0 & 0 & 1 \end{bmatrix} \tag{4}$$

where $\mathbf{v}$ denotes the velocity vector, $\dot{\boldsymbol{\eta}}$ is the position vector and $\boldsymbol{R}(\boldsymbol{\eta})$ is the rotation matrix that maps vectors from a body fixed frame to an inertial frame.

The planar dynamic model of the surface vehicle can be described as follows:

$$M\dot{v} + C(v)v + D(v)v = f \tag{5}$$

where M denotes the mass matrix, which includes body and added masses, C is the Coriolis and centripetal matrix, D is the matrix of the damping coefficients and f is the control force and moment. Since the propulsion system of the DW-uBoat is a differential thruster type, according to Sonnenburg et al. (2013), f can be described as follows:

$$f = \begin{bmatrix} \tau_X \\ \tau_Y \\ \tau_N \end{bmatrix} = \begin{bmatrix} T_{port} + T_{stbd} \\ 0 \\ (T_{port} - T_{stbd})B/2 \end{bmatrix} \tag{6}$$

where T_{port} and T_{stbd} denote the thrust force of the port and starboard side thrusters, respectively, and B is the beam of the DW-uBoat.

Therefore, the three-degrees-of-freedom nonlinear dynamic motion equation can be shown as follows:

$$\dot{u} = \frac{X_{|u|u}}{m - X_{\dot{u}}}|u|u + \frac{X_u}{m - X_{\dot{u}}}u + \frac{1}{m - X_{\dot{u}}}\tau_X \tag{7}$$

$$\begin{aligned} \dot{v} = {} & \tfrac{I_z - N_{\dot{r}}}{\nabla}((-Y_v - Y_{|v|v}|v| - Y_{|r|v}|r|)v + ((m - X_{\dot{u}})u_0 - Y_r - Y_{|v|r}|v| - \\ & Y_{|r|r}|r|)r + \tau_Y) - \tfrac{mx_G - Y_{\dot{r}}}{\nabla}((-(m - X_{\dot{u}})u_0 - N_v - N_{|v|v}|v| - N_{|r|v}|r|)v \\ & + (-N_r - N_{|v|r}|v| - N_{|r|r}|r|)r + \tau_N) \end{aligned} \tag{8}$$

$$\begin{aligned} \dot{r} = {} & \tfrac{N_{\dot{v}} - mx_G}{\nabla}((-Y_v - Y_{|v|v}|v| - Y_{|r|v}|r|)v + (m - X_{\dot{u}})u_0 - Y_r - Y_{|v|r}|v| \\ & - Y_{|r|r}|r|)r + \tau_Y) + \tfrac{m - Y_{\dot{v}}}{\nabla}((-(m - X_{\dot{u}})u_0 - N_v - N_{|v|v}|v| - N_{|r|v}|r|)v \\ & + (-N_r - N_{|v|r}|v| - N_{|r|r}|r|)r + \tau_N) \end{aligned} \tag{9}$$

where $X_{(\cdot)}$, $Y_{(\cdot)}$ and $N_{(\cdot)}$ denote the constant hydrodynamic coefficients, which are the partial derivatives of the surge, sway force and yaw moments, respectively. The detailed hydrodynamic coefficients can be viewed as follow:

$$\begin{cases} \left[X_{|u|u}, X_u, X_{\dot{u}}\right] \\ \left[Y_{\dot{v}}, Y_{\dot{r}}, Y_{|v|v}, Y_v, Y_{|r|v}, Y_r, Y_{|v|r}, Y_{|r|r}\right] \\ \left[N_{\dot{v}}, N_{\dot{r}}, N_v, N_{|v|v}, N_{|r|v}, N_r, N_{|v|r}, N_{|r|r}\right] \end{cases} \tag{10}$$

3. PINNs

3.1. ANNs

ANNs have been a research hotspot in the field of AI since the 1980s. They abstract the neural network of the human brain from the perspective of information processing to establish simple models and forms of different networks according to different connection modes. Mathematically, ANNs can be described as direct graphs composed of a group of vertices representing neurons and a group of edges representing links. There are many variants of neural networks, such as feedforward, conventional and recurrent neural networks, that perform various increasingly complex tasks. In this study, the feedforward neural network, which is also referred to as a multilayer perceptron, is adopted to solve the problem.

The simplest structure of a multilayer perceptron contains a single input layer, single hidden layer and single output layer and can be viewed in Figure 3. In this case, the operating principle of the single cell body can be described in Figure 4 and the output in the hidden layer is calculated as follows:

$$y_j = f_j\left(b_j + \sum_{i=1}^{d} w_{i,j}x_i\right) \tag{11}$$

where $w_{i,j}$ denotes the weights from the input layer to the hidden layer, b_j represents the threshold values and f_i is the nonlinear activation function. The activation function can endow the neurons with the ability to solve nonlinear problems, which means that the neural network decodes any nonlinear function arbitrarily. The commonly adopted activation functions include sigmoid, Tanh, ReLU and Leaky ReLU functions, as demonstrated in Figure 5.

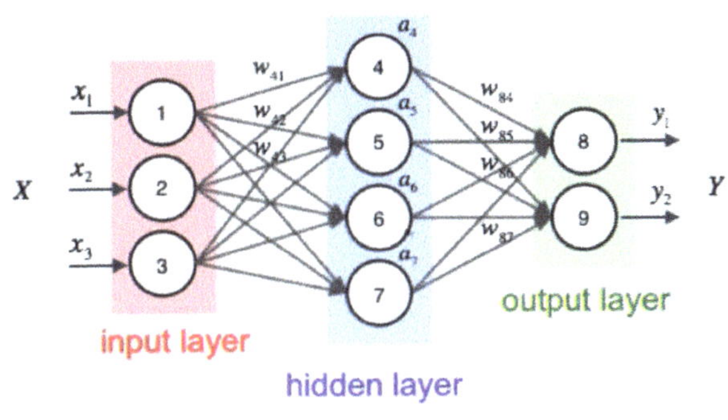

Figure 3. Structure of the deep neural network.

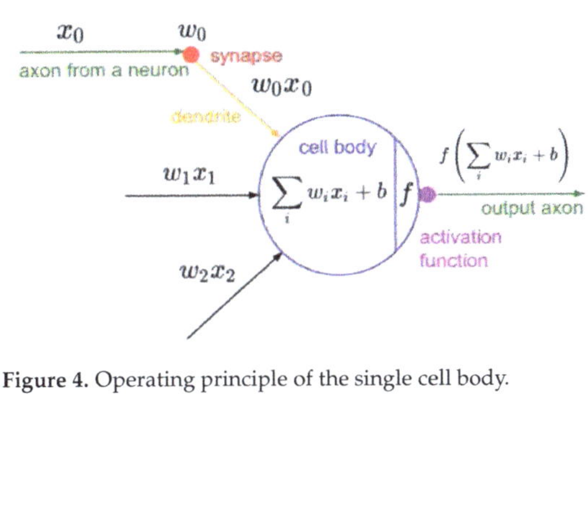

Figure 4. Operating principle of the single cell body.

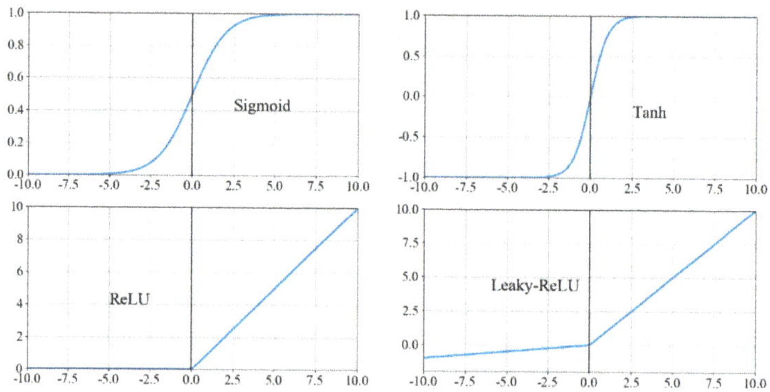

Figure 5. Commonly used activation functions.

In addition, it is also possible to incorporate additional hidden layers between the input and output layers. Therefore, the output in another hidden layer can be expressed as follows:

$$y_k = f_k\left[b_k^{(2)} + \sum_{i=1}^{m_2} w_{j,k}^{(2)} \cdot f_j(b_j^{(1)} + \sum_{i=1}^{m_1} w_{i,j}^{(1)} x_j)\right] \tag{12}$$

An extension of this process as a simple application of the chain rule can be described as follows:

$$y(x) = f_d(\cdots f_2(f_1(x))) \tag{13}$$

"Deep" neural networks mean that the structure of the neural network includes many hidden layers between the input and output layers. The advantages of increasing the amount of hidden layers can reduce the computing cost in some practices and can solve more complex nonlinear problems with less data. However, Hagan et al. (1994) proposed that the performance of a neural network is weakened with increasing hidden layers [17]. He et al. (2016) also put forward that the process from input to output is almost irreversible (information loss) due to the existence of the nonlinear activation function [18]. It is difficult to obtain the original input from the output. Therefore, it is significant that the neural network needs to benefit from the identity mapping. The residual neural network is proposed to ensure that the internal structure of the model has an identity mapping capability in order to ensure that network will not degenerate in the stacking of additional hidden layers.

3.2. PINN Method

Raissi et al. (2017) first proposed the PINN to solve the data-driven solution and discovery of partial differential equations [19]. A PINN is a neural network that is trained to solve supervised learning tasks while respecting any given law of physics described by general nonlinear partial differential equations.

We assume a simple form of the parametrized and nonlinear partial differential equation, as follows:

$$u_t + N(u;\lambda) = 0, \quad x \in \Omega, \quad t \in [0,T] \tag{14}$$

where $u(t,x)$ denotes the latent solution of the partial differential equation and $N(u;\lambda)$ is a nonlinear operator parametrized by λ. Equation (13) represents a wide range of problems in mathematical physics, including conservation laws, diffusion processes, advection-diffusion-reaction systems and kinetic equations.

The physical problem can then be cast into the form of Equation (14) and its solution can be calculated by minimizing the loss function of a neural network. The target of the loss function is the mean squared error loss, as expressed in Equation (15), which contains two parts. The loss MSE_u denotes the loss function of the initial and boundary conditions, while MSE_f enforces the structure imposed by a target partial differential equation at a finite set of collocation points.

$$f \equiv u_t + N(u;\lambda) \tag{15}$$

$$MSE = MSE_u + MSE_f \tag{16}$$

$$MSE_u = \frac{1}{N_u}\sum_{i=1}^{N_u}\left|u\left(t_u^i, x_u^i\right) - u^i\right|^2 \tag{17}$$

$$MSE_f = \frac{1}{N_f}\sum_{i=1}^{N_f}\left|f\left(t_f^i, x_f^i\right)\right|^2 \tag{18}$$

where $\{t_u^i, x_u^i, u^i\}_{i=1}^{N_u}$ denotes the initial and boundary training data on solution of the equation and $\left\{t_f^i, x_f^i\right\}_{i=1}^{N_f}$ are the collocation points for $f(t,x)$.

3.3. *PINN for Solving Dynamic Models of the USV*

The process of PINN solving the dynamic models of the USV can be viewed in Figure 6. The time t is selected as input, the velocities u, v and r are deemed as outputs and the layers and neurons in each layer between input and output, which constitute the fully connected-NN structure. The dynamic models of USV should be transformed as Equation (14). For illustration purposes only, the structure of a network with three hidden layers and five neurons per hidden layer is demonstrated in Figure 6. Actually, the structure of the network includes 6 layers with 8 neurons in each layer. The dynamic models are then embedded into the loss function. The automatic differentiation technique is adopted to compute the physical-based loss function and the velocity u on the training data is calculated by minimizing the loss function. In addition, I in the green box denotes the identity operator, while the ∂_t is the differential operator, which can be explained as an activation operator. The speed and steering models are embedded into the loss function and all the hydrodynamic coefficients are determined [20]. In order to reduce the error of the loss function, the Adam optimizer is utilized to optimize the target function. The Adam optimizer can constantly adjust the learning rates with the situation changes in the learning process. The 'Xavier' method is designed to decide the initial weights and biases which can ensure faster convergence of the neural network. A residual neural network is added in the FCNN to avoid gradient explosion and/or gradient disappearance.

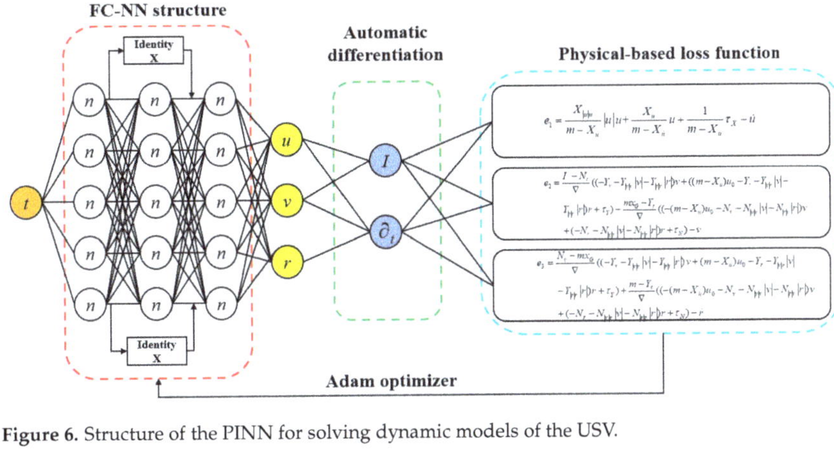

Figure 6. Structure of the PINN for solving dynamic models of the USV.

4. Results

4.1. Data Preprocessing

The experiment was operated in the Qing Huai River in Nanjing, as shown in Figure 7, with constant wind and current always present at this location. It is noted that the wind and current is not strong and has a low influence on the performance of the USV. The DW-uBoat voyages in the Qing Huai River can be seen in Figure 8. Two straight lines with varying rpm of differential thrusters are used to estimate the speed model and two different zigzag trajectories are designed for training and testing the steering model. The data collected by the receiver per 0.25 s and the 1000 s′ data is adopted to train the dynamic model of the USV. Furthermore, the normalization technique is used to guarantee the accurate forecast results due to the large range among each dimension of input data. The min-max normalization is selected and can be expressed as:

$$x' = \frac{x - x_{max}}{x_{max} - x_{min}} \tag{19}$$

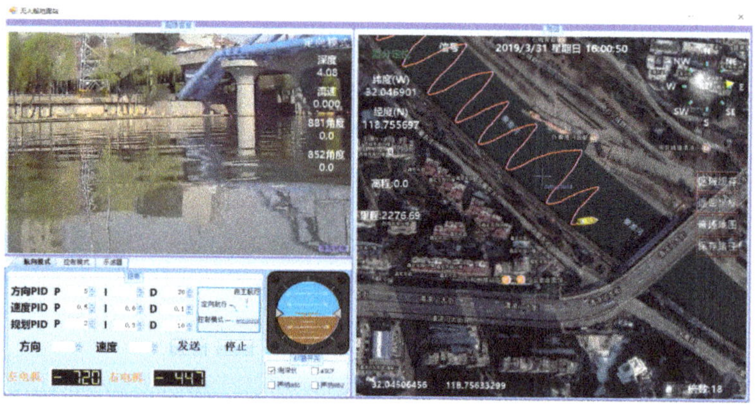

Figure 7. Qing Huai River in Google Maps.

Figure 8. DW-uBoat voyage on the Qing Huai River.

4.2. Identified Results

In this section, the number of training data is 4000 and the number of testing data is 4000. The results can be viewed in Figure 9. The training and testing data are obtained by different straight-line and zigzag trajectories. The first picture represents the predicted and true results of the straight-line motions while the last two pictures represent the predicted and true results of the zigzag trajectories. Left of the red line is the training set and right of the red line is the testing set. The blue lines represent the experiment results and yellow dotted lines represent the results predicted by the PINN. It is obvious that the PINN has a strong ability for reconstructing the dynamics of the USV, since the mean square errors of the surge and sway velocities and rotation speed are small. In addition, the PINN also has a good generalization ability in the testing set, meaning that it can accurately predict the unknown trajectory. It is noteworthy that the number of training data is not large, which means we can obtain the dynamic model of the USV using a small dataset and high convergence speed.

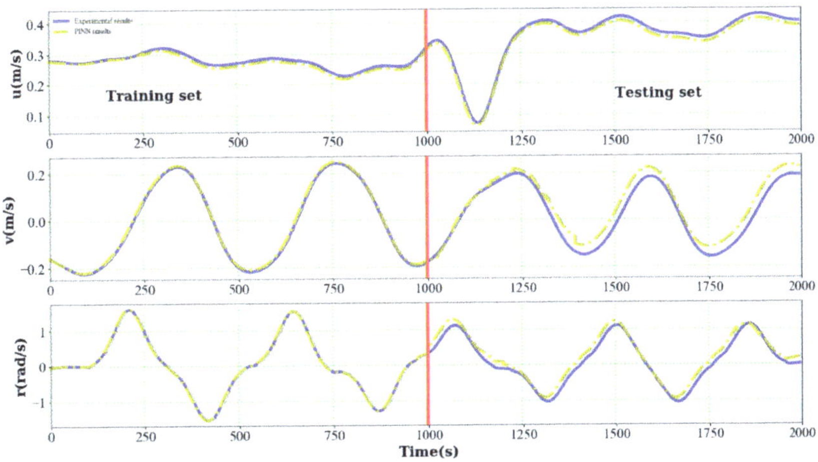

Figure 9. Observed and predicted surge, sway and rotation velocities in training and test sets.

4.3. PINN Versus Traditional Neural Network

In this section, the PINN method is compared with a tradition neural network. We select the same structure of the neural network (fully-connected with four layers and ten neurons in each layer) and training data. The mean square errors are utilized to test the performance of the PINN and traditional neural network. Table 1 shows the identification results of the hydrodynamic coefficients. A comparison between the PINN and traditional neural network in the testing set under different numbers of samples is given in Table 2. It is obvious that the PINN has a better generalization ability for predicting the dynamic models of the USV. The PINN method is lower than the traditional neural network method in order of magnitude. The PINN can also predict the dynamic models under low numbers of training samples and a simple neural structure.

Table 1. The results of system identification.

u	Value	v	Value	r	Value
$X_{\|u\|u}$	−0.002	$Y_{\dot{v}}$	0.00237	$N_{\dot{v}}$	0.000849
X_u	0.423	$Y_{\dot{r}}$	0.006043	$N_{\dot{r}}$	−0.00517
$X_{\dot{u}}$	0.0992	$Y_{\|v\|v}$	-0.54×10^{-5}	N_v	0.423456
		Y_v	−0.00015	$N_{\|v\|v}$	-2.73×10^{-7}
		$Y_{\|r\|v}$	0.001508	$N_{\|r\|v}$	0.003286
		Y_r	0.00211	N_r	0.003142
		$Y_{\|v\|r}$	−0.00038	$N_{\|v\|r}$	0.000777
		$Y_{\|r\|r}$	0.00265	$N_{\|r\|r}$	0.000105

Table 2. Comparison between MSE of PINN and traditional NN in the testing set under different numbers of training samples.

	PINN			Traditional NN		
	u	v	r	u	V	r
1000	1.3×10^{-4}	7.2×10^{-4}	2.8×10^{-2}	7.2×10^{-3}	9.3×10^{-3}	1.8×10^{-1}
2000	0.4×10^{-4}	4.4×10^{-4}	3.3×10^{-3}	6.6×10^{-3}	6.9×10^{-3}	0.9×10^{-1}
3000	1.1×10^{-5}	0.6×10^{-4}	1.8×10^{-3}	2.4×10^{-3}	4.2×10^{-3}	1.3×10^{-2}
4000	1.3×10^{-5}	1.1×10^{-5}	1.2×10^{-3}	9.1×10^{-4}	0.7×10^{-4}	0.4×10^{-2}
5000	1.2×10^{-5}	1.1×10^{-5}	1.3×10^{-3}	7.7×10^{-4}	6.1×10^{-5}	0.6×10^{-2}

5. Conclusions

In this study, the PINN method was first proposed to identify the dynamic models of a USV. Zigzag and straight-line tests were carried out in the Qing Huai river to obtain the data. The physics-driven deep learning method, instead of the data-driven deep learning method, was utilized to obtain the USV dynamics. The speed model and steering model of the USV were embedded into the loss function which can guarantee the loss function, including the physical information. The results show that the PINN has a better generalization ability in predicting the sway and surge velocities and rotation speed under a small number of training samples and the simple structured neural network compared with the traditional neural network. There is a reasonable prospect that the PINN method can replace traditional deep learning for the identification of the dynamic models of unmanned surface vehicles. In the future, the model of the USV with wind, waves and other disturbances will be embedded into the PINN and identified.

Author Contributions: Conceptualization, C.C.; methodology, P.-F.X.; validation, C.-B.H.; formal analysis, C.C.; data curation, C.-B.H.; writing—original draft preparation, C.C.; writing—review and editing, H.-X.C.; supervision, P.-F.X.; project administration, T.G.; funding acquisition, T.G. All authors have read and agreed to the published version of the manuscript.

Funding: National Natural Science Foundation of China (52071131), the Marine Science and Technology Innovation Project of Jiangsu Province (HY2018-15), the National Key Research and Development Program of China (2018YFF0215005), and the China Postdoctoral Science Foundation (2018M640390).

Data Availability Statement: Not applicable.

Acknowledgments: Thanks to Yalin Shen, Yanxu Ding for their preliminary preparations for this article.

Conflicts of Interest: The authors declare no conflict of interest.

References

1. Fang, Y.; Pang, M.Y.; Wang, B. A course control system of unmanned surface vehicle (USV) using back-propagation neural network (BPNN) and artificial bee colony (ABC) algorithm. In Proceedings of the 8th International Conference on Advances in Information Technology, IAIT2016, Macau, China, 19–22 December 2017.
2. Fossen, T.I. *Marine Control Systems: Guidance, Navigation and Control of Ships, Rigs and Underwater Vehicles*; Marine Cybernetics: Trondheim, Norway, 2002.
3. Woo, T.; Park, J.Y.; Yu, C.; Kim, N. Dynamic model identification of unmanned surface vehicles using deep learning network. *Appl. Ocean. Res.* **2018**, *78*, 123–133. [CrossRef]
4. Sonnenburg, C.R.; Woolsey, C.A. Modeling, identification, and control of an unmanned surface vehicle. *J. Field Rob.* **2013**, *30*, 371–398. [CrossRef]
5. Nagumo, J.I.; Noda, A. A learning method for system identification. *IEEE Trans. Autom. Control. AC* **1967**, *12*, 282–287. [CrossRef]
6. Holzhuter, T. Robust identification scheme in an adaptive track controller for ships. In Proceedings of the 3rd IFAC Symposium on Adaptive System in Control and Signal Processing, Glasgow, UK, 19–21 April 1989; pp. 118–123.
7. Kallstrom, C.G.; Astrom, K.J. Experiences of system identification applied to ship steering. *Automatica* **1981**, *17*, 187–198. [CrossRef]
8. Yoon, H.K.; Rhee, K.P. Identification of hydrodynamic coefficients in ship maneuvering equations of motion by estimation-before-modeling technique. *Ocean Eng.* **2003**, *30*, 2379–2404. [CrossRef]

9. Shin, J.; Kwak, D.J.; Lee, Y.-I. Adaptive path following control for an unmanned surface vessel using an identified dynamic model. *IEEE/ASME Trans. Mechatron.* **2017**, *22*, 1143–1153. [CrossRef]
10. Selvam, R.P.; Bhattacharyya, S.; Haddara, M. A frequency domain system identification method for linear ship maneuvering. *Int. Shipbuild. Prog.* **2005**, *52*, 5–27.
11. Alattas, K.A.; Mobayen, A.; Din, S.U.; Asad, J.H.; Fekih, A.; Assawinchaichote, W.; Vu, M.T. Design of a Non-Singular Adaptive Integral-Type Finite Time Tracking Control for Nonlinear Systems with External Disturbances. *IEEE Access* **2020**, *9*, 102091–102103. [CrossRef]
12. Vu, M.T.; Thanh, H.N.N.; Huynh, T.T.; Do, Q.T.; Do, T.D.; Hoang, Q.D.; Le, T.H. Station-Keeping Control of a Hovering Over-Actuated Autonomous Underwater Vehicle Under Ocean Current Effects and Model Uncertainties in Horizontal Plane. *IEEE Access* **2020**, *9*, 6855–6867. [CrossRef]
13. Thanh, H.L.N.N.; Vu, M.T.; Mung, N.X.; Nguyen, N.P.; Phuong, N.T. Perturbation Observer-Based Robust Control Using a Multiple Sliding Surfaces for Nonlinear Systems with Influences of Matched and Unmatched Uncertainties. *Mathematics* **2020**, *8*, 1371. [CrossRef]
14. Rajesh, G.; Bhattacharyya, S. System identification for nonlinear maneuvering of large tankers using artificial neural network. *Appl. Ocean Res.* **2008**, *30*, 256–263. [CrossRef]
15. Oskin, D.A.; Dyda, A.A.; Markin, V.E. Neural Network Identification of Marine Ship Dynamics. In Proceedings of the 9th IFAC Conference on Control Applications in Marine Systems. The International Federation of Automatic Control, Osaka, Japan, 17–20 September 2013.
16. Pan, C.Z.; Lai, X.Z.; Yang, S.X.; Wu, M. An efficient neural network approach to tracking control of an autonomous surface vehicle with unknown dynamics. *Expert Syst. Appl.* **2013**, *40*, 1629–1635. [CrossRef]
17. Hagan, M.T.; Menhaj, M.B. Training feedforward networks with the Marquardt algorithm. *IEEE Trans. Neural Netw.* **1994**, *5*, 989–993. [CrossRef] [PubMed]
18. He, K.; Zhang, X.; Ren, S.; Sun, J. Deep residual learning for image recognition. In Proceedings of the IEEE Conference on Computer Vision and Pattern Recognition, Las Vegas, NV, USA, 27–30 June 2016; pp. 770–778.
19. Raissi, M.; Perdikaris, P.; Karniadakis, G.E. Physics Informed Deep Learning (Part I): Data-driven Solutions of Nonlinear Partial Differential Equations. *arXiv* **2017**, arXiv:1711.10671.
20. Xu, P.F.; Cheng, C.; Cheng, H.X.; Shen, Y.L.; Ding, Y.X. Identification-based 3 DOF model of unmanned surface vehicle using support vector machines enhanced by cuckoo search algorithm. *Ocean Eng.* **2020**, *197*, 1–11. [CrossRef]

Journal of
Marine Science and Engineering

MDPI

Article

Adaptive Integral Sliding Mode Based Course Keeping Control of Unmanned Surface Vehicle

José Antonio González-Prieto [1,*], Carlos Pérez-Collazo [1,*] and Yogang Singh [2]

1 Defense University Center, Spanish Naval Academy, Plaza de España, s/n, 36920 Marin, Galicia, Spain
2 Industrial Systems Engineering (ISyE), FlandersMake@Ghent University, Graaf Karel de Goedelaan 2B, Geb.A, 8500 Kortrijk, Belgium; yogang.singh@ugent.be
* Correspondence: jose.gonzalez@cud.uvigo.es (J.A.G.-P.); carlos.perez.collazo@cud.uvigo.es (C.P.-C.)

Abstract: This paper investigates the course keeping control problem for an unmanned surface vehicle (USV) in the presence of unknown disturbances and system uncertainties. The simulation study combines two different types of sliding mode surface based control approaches due to its precise tracking and robustness against disturbances and uncertainty. Firstly, an adaptive linear sliding mode surface algorithm is applied, to keep the yaw error within the desired boundaries and then an adaptive integral non-linear sliding mode surface is explored to keep an account of the sliding mode condition. Additionally, a method to reconfigure the input parameters in order to keep settling time, yaw rate restriction and desired precision within boundary conditions is presented. The main strengths of proposed approach is simplicity, robustness with respect to external disturbances and high adaptability to static and dynamics reference courses without the need of parameter reconfiguration.

Keywords: unmanned surface vehicle; Guidance, Navigation and Control; course keeping; adaptive sliding mode

Citation: González-Prieto, J.A.; Pérez-Collazo, C.; Singh, Y. Adaptive Integral Sliding Mode Based Course Keeping Control of Unmanned Surface Vehicle. *J. Mar. Sci. Eng.* **2022**, *10*, 68. https://doi.org/10.3390/jmse10010068

Academic Editor: Weicheng Cui

Received: 22 November 2021
Accepted: 30 December 2021
Published: 6 January 2022

1. Introduction

With the growing advancement in the sensor technology and navigation aids, USVs are becoming a popular tool in maritime domain for several applications ranging from environmental monitoring, military surveillance to scientific surveying, and data collection. Mission oriented approach of USVs subject them to several types of maritime environment comprising of wind, wave, and sea surface currents leading to requirement of designing and developing several autonomy levels for successful operation. Henceforth, design and development of approaches for Guidance, Navigation, and Control (GNC) of a USV is an important research area for constructing operational and tactical approaches for seven different operational autonomy level of USVs as described by International Maritime Organisation (IMO).

Guidance and control of USV plays an important role in motion control system to manipulate the forces to enable a USV to follow a desired path whilst maintaining the stability. Three approaches, namely, waypoint control, path following control and trajectory tracking are generally considered in the domain of marine robotics to enable a USV to follow a designated path [1,2]:

- **Waypoint control:** In this strategy, Line of Sight (LOS) based approach is adopted to follow a certain waypoints, generated heuristically, in the required maritime environment.
- **Path following control:** In this strategy, a path generated through path planning algorithms is used as a reference, to be followed with no temporal constraints. Here, USV should converge and follow the desired path without any time constraints and simultaneously satisfies its assigned velocity profile.
- **Trajectory tracking:** In this strategy, temporal constraints are enforced upon the path generated using path planners. This is predominantly used with fully actuated marine vehicles reasoned with better maneuvering capabilities.

Sailing conditions and unpredictability of environmental disturbances can have a significant impact on the ship's dynamics. It is therefore necessary to develop a nonlinear controller that overcomes unknown disturbances and ensures robustness. As long as parameter uncertainties and unknown bounded disturbances remain, the adaptive method is likely to remain a superior approach. It is intended that vessel steering autopilots will force the ship to follow a predetermined course with a fixed speed by controlling the rudder angle, creating a course keeping problem that the current study is attempting to resolve.

1.1. State of the Art

The problem of **course keeping control** is highly non-linear in nature and has been studied from a perspective of observed disturbance control using sliding mode control (SMC) approach. The SMC problem for USVs, subjected to, higher order non linear operational disturbances, have been studied with varying control approaches like sliding mode [3–6]; fuzzy sliding mode [7]; proportional derivative fuzzy [8]; backstepping [9–12]; backstepping with adaptive radial basis function neural network [13]; sine function-based non-linear feedback [14]; hyperbolic tangent based nonlinear control [15]; sigmoid based nonlinear control [16]; function adaptive neural path following control [17]; model predictive control [18,19]; event-triggered control approach [20] and non-linear feedback power functions [21].

In order to make control robust to disturbances and uncertainties, several approaches has been proposed in the SMC literature, see [22–32]. Some proposals of advanced sliding manifolds include recursive nonlinear sliding manifolds [33–35], adaptive integral sliding mode approach [36–38], non linear full order dynamics [39,40], sliding surfaces with adaptive damping parameters [41–43] and, in the last years, a vast collection of homogeneity based works, see [44] for instance. Applications of the properties of homogeneous systems is an important field of study in the current development of analysis and design of nonlinear controllers and observers. Homogeneity simplifies analysis and design of nonlinear control systems since the homogeneous vector fields have many properties similar to linear one and provides solutions with finite-time and fixed-time stability.

The dynamics generated by an homogeneous controller can be seen as a lineal dynamic system with an adaptive gain that grows to ∞ as $|x(t)| \to 0$, generating the well know singularity at the origin which is undesired for real applications. Nevertheless, as commented in [45], the practical implementation of homogeneous dynamics system designed in the continuous time domain prevents the use of explicit *Euler* discretization scheme to achieve a mere copy of the continuous time approach due to its simplicity. This type of discretization is considered inappropriate, especially when set-valued functions has to be considering, causing numerical chattering and sensitivity to the gains. As a result, without addressing the discretization issue, any comparison between homogeneous based solutions and other types of proposals may potentially lead to unfair conclusions.

Based on the aforementioned results, in order to keep the discretization process simple, an adaptive lineal sliding mode surface law, that includes a nested integral sliding surface is introduced in this work. In this case, the dynamics flows with adaptive and finite damper gain, avoiding the effects of the peaking transient response inherent to linear systems and allowing fast responses at steady state, approximating the behaviour obtained with homogeneous solutions.

1.2. Major Contributions

The paper makes following contributions to the current state of existing approaches to SMC techniques for USVs:

- A number of simulation studies in the manuscript demonstrate that the proposed adaptive control approach can be reconfigured for various input trajectories and marine environmental disturbances, without requiring parametric adjustment.
- The cut-off frequency of the system response is an indication of the bound to be assigned to the disturbance derivative in the algorithm. This relationship is based on low-pass filtering properties associated with the second order adaptive linear dynamics generated

at the sliding variable. As a result, frequencies over ω_c do not affect the sliding variable response. In practice, this feature offers some advantages when estimating the maximum value of the disturbance derivative is a challenging task.
- The proposed adaptive profile generates low/high gains based on the absolute error. As a result, the control input is not saturated when there is a large error (gain is small) and the response at steady state becomes fast disturbance compensation (gain is large).
- Based on the adaptive placement of two poles relating to a second order dynamical system with critical damping, we can generate an overdamped response that avoids the occurrence of considerable overshoots.
- By avoiding the need of the derivative of the fractional power terms with respect to time, the singularity problem associated with terminal sliding mode solutions can be avoided. Thus, the high sensitive performance around the equilibrium point generated by set value or fractional order functions can be reduced.

This paper has been structured as follows. First, in Section 2 we present the nonlinear dynamic model of the course keeping problem, the desired objectives to be achieved and a theoretical stability tool that is used in the posterior analysis of the control algorithm. Then, Section 3 describes the proposed adaptive integral sliding mode (AISM) algorithm. Results from numerical simulations are then presented and discussed in Section 4. Finally, conclusions are drawn in Section 5.

2. Problem Statement

The motion of the USV is shown in Figure 1, where a six degrees of freedom (DOF) model is presented. The earth fixed **Oo** is an inertial reference frame fixed to the earth's surface and the body fixed with origin **O** is a moving coordinate frame that it is fixed to the craft as in given in [1]. It is assumed an homogeneous mass distributed and xz-plane symmetrical, such that origin of the body fixed reference frame is chosen to be coincident with the center of gravity.

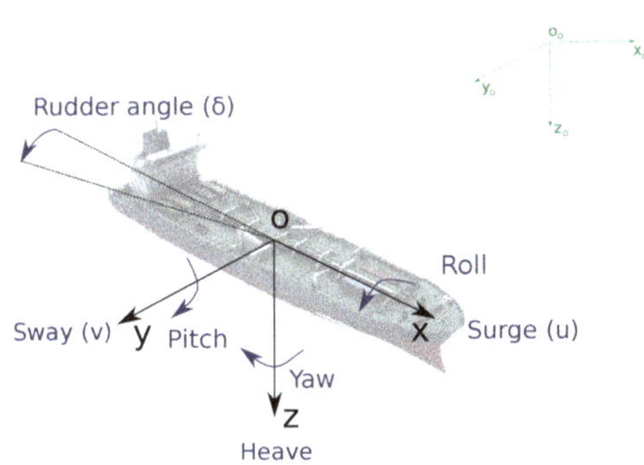

Figure 1. 6 DOF motion representation with North-East-Down coordinate system (green) and body fixed reference frame (black).

If we consider the course keeping problem the dynamics of heave, roll, and pitch can be neglected, so that the reduced model dynamics are given as

$$m(\dot{u} - vr - x_c r^2) = X \tag{1}$$

$$m(\dot{v} + ur + x_c \dot{r}) = Y \tag{2}$$

$$I_z \dot{r} + m x_c(\dot{v} + ur) = N \tag{3}$$

where m is the mass, u is the surge velocity, v is the sway velocity, r is the yaw rate, I_z is the rotational inertia with respect to z axis, x_c is the x coordinate of the vehicle center in the fixed body reference frame and X, Y and N are the external forces and moments with respect to the surge, sway, and yaw, respectively.

Assumption of constant forward speed and using the ship's *Norrbin* nonlinear mathematical model, see [46], implies that the steering equations of motion can be obtained as

$$\begin{aligned} \dot{\psi}(t) &= r(t) \\ \dot{r}(t) &= f(r) + g\delta(t) + d(t) \end{aligned} \tag{4}$$

where, $\psi(t)$ is the yaw (orientation) angle, $r(t)$ is the yaw rate, $\delta(t)$ is the rudder angle (the control variable to be designed) and $d(t)$ is an unknown term to be compensated that includes parametric uncertainty and external disturbances (wind, waves, mobile loads). The dynamics functions are given as

$$\begin{aligned} g &= -\frac{K}{T} \\ f(r) &= -\frac{K}{T}H(r) \\ H(r) &= a_1 r + a_2 r^3 \end{aligned} \tag{5}$$

where (K, T) are hydrodynamic coefficients and (a_1, a_2) are *Norrbin* coefficients.

In the course keeping problem it is required that the yaw angle ψ follows a reference angle ψ_r by means of the design of the rudder control signal $\delta(t)$. The following assumptions are taking account in this work.

Assumption 1. *$d(t)$ in (4) satisfies the following restriction*

$$|d(t)| \leq D$$

with $D > 0$ a positive unknown real number.

Assumption 2. *$d(t)$ in (4) satisfies the following restriction*

$$|\dot{d}(t)| \leq \Delta$$

with $\Delta > 0$ a positive known real number.

Assumption 3. *Henceforth, it is assumed that a reference yaw establish the desired input to be tracked, which can be obtained by means of path planning algorithms, that account for different environment constraints as in [47–50]. A dynamic reference model is used, in this work, to generate the desired course $(\psi_r(t), \dot{\psi}_r(t), \ddot{\psi}_r(t))$.*

The objective is to design a control law that creates overdamped responses with minimal overshooting (undershooting) and robustness properties for response of the yaw error, which is defined as

$$e(t) = \psi(t) - \psi_r(t) \tag{6}$$

In order to check the control performance of the proposed controller for the course keeping problem, we consider the following performance analysis indices mentioned in [10,12].

$$MAE = \frac{1}{t_\infty - t_0}\int_{t_0}^{\infty} |e(t)|dt \tag{7}$$

$$MIA = \frac{1}{t_\infty - t_0}\int_{t_0}^{\infty} |\delta(t)|dt \tag{8}$$

$$MTV = \frac{1}{t_\infty - t_0}\int_{t_0}^{\infty} |\delta(t) - \delta(t-\tau)|dt \tag{9}$$

where τ is the sampling time used in the simulation.

Furthermore, to check out the robustness properties of the solution, we compare the results with the algorithms proposed in [10,12] applying the following conditions:

- As in [10,12], we test two problems that uses two different types of reference input signals: step and sinusoidal.
- The tests includes results without disturbances ($d(t) = 0$) and with disturbances ($d(t) \neq 0$).
- The algorithm parameters are configured in the case of the step input reference without disturbances, such that all solutions provide the same value of the MIA index at the end of the test time.
- After that, the algorithms parameters are fixed and tested in the case of step with disturbances and in the case of the sinusoidal input reference. In this way we check the robustness of the solutions with respect to its capacity of adaptation to different scenarios from a specific parameter configuration.

The following theorem is introduced in order to analyse the stability properties of the AISM proposed solution.

Theorem 1. *Consider the following cascade system*

$$\dot{z}_1 = f_1(t, z_1) + g_1(t, z_1, z_2)z_2 \tag{10}$$

$$\dot{z}_2 = f_2(t, z_2) \tag{11}$$

where $z_1 \in \mathbb{R}^n$, $z_2 \in \mathbb{R}^m$, $f_1(t, z_1)$ is continuously differentiable in (t, z_1), and $f_2(t, z_1)$ and $g_1(t, z_1, z_2)$ are continuous and locally Lipschitz in z_2 and (z_1, z_2), respectively.

The dynamics of (10) when $z_2 = 0$ are

$$\dot{z}_1 = f_1(t, z_1) \tag{12}$$

If systems (12) and (11) are globally uniformly asymptotically stable (GUAS) and we know a C^1 Lyapunov function $V(t, z_1)$, two class-K_∞ functions ϕ_1 and ϕ_2, a class-K ϕ_3 function and a positive semidefinite function $W(z_1)$ such that

$$\phi_1(||z_1||) \leq V(t, z_1) \leq \phi_2(||z_1||) \tag{13}$$

$$\frac{\partial V}{\partial t} + \frac{\partial V}{\partial z_1} f_1(t, z_1) \leq -W(z_1) \tag{14}$$

$$||\frac{\partial V}{\partial z_1}|| \leq \phi_3 \tag{15}$$

Besides, for each fixed z_2 there exists a continuous function $\zeta : \mathbb{R}^+ \to \mathbb{R}^+$ such that

$$\lim_{s \to \infty} \zeta(s) = 0 \tag{16}$$

$$||\frac{\partial V}{\partial z_1} g_1(t, z_1, z_2)|| \leq \zeta(||z_1||)W(z_1) \tag{17}$$

Then we can conclude that the cascade system (10) and (11) is GUAS.

Proof. See [51]. □

3. Adaptive Integral Sliding Mode Surface Control Design

Derivation of $e(t)$ in (6) leads to

$$\dot{e}(t) = r(t) - \dot{\psi}_r(t) \tag{18}$$

An adaptive sliding surface $s(t)$ variable is defined as

$$s(t) = \dot{e}(t) + \lambda(e)e(t) \tag{19}$$

with $\lambda(e)$ a real positive time varying parameter.

Consider the integral term $\bar{s}(t)$

$$\bar{s}(t) = \int_0^t s(t)dt \tag{20}$$

Let us choose the control law as

$$\delta(t) = \frac{1}{g}(-f(r) + \ddot{\psi}_r(t) - \lambda(e)\dot{e}(t) - \dot{\lambda}(e)e(t) - \alpha(s,\bar{s})s(t) - \gamma(e)\bar{s}(t)) \tag{21}$$

with $\lambda(e)$ defined as

$$\lambda(e) = \max(\lambda_{min}, \lambda_{max} - (\frac{\lambda_{max} - \lambda_{min}}{|e(0)|})|e(t)|), \tag{22}$$

the variable $z(t)$, related a new sliding surface, defined as

$$z(t) = s(t) + \frac{\alpha}{2}\bar{s}(t) \tag{23}$$

and the parameters $\alpha(s,\bar{s})$, $\gamma(\alpha)$ given as

$$\dot{\alpha}(s,\bar{s}) = \begin{cases} \kappa|z|^{\zeta}\,\text{sign}(z)\,\text{sign}(s) & \text{if } |z| > \mu \wedge \alpha_{min} < \alpha < \alpha_{max} \\ 0 & \text{otherwise} \end{cases} \tag{24}$$

$$\gamma(\alpha) = \frac{\alpha^2}{4} \tag{25}$$

with $\zeta(e)$ defined as

$$\zeta(e) = (\frac{\zeta_{max} - \zeta_{min}}{|e(0)|})|e(t)| + \zeta_{min} \tag{26}$$

and μ

$$\mu = \sqrt[\zeta+1]{\frac{\Delta}{\kappa}} \tag{27}$$

Derivation of $\gamma(\alpha)$ and $\lambda(e)$ are given as

$$\dot{\gamma}(\alpha) = \frac{\alpha}{2}\dot{\alpha} \tag{28}$$

$$\dot{\lambda}(e) = \begin{cases} -(\frac{\lambda_{max}-\lambda_{min}}{|e(0)|})\text{sign}(e(t))\dot{e}(t) & \text{if } \lambda > \lambda_{min} \\ 0 & \text{if } \lambda \leq \lambda_{min} \end{cases} \tag{29}$$

The control algorithm is designed by an appropriate selection of the parameters α_{max}, α_{min}, λ_{max}, λ_{min}, $\alpha(0)$, κ, ζ_{max} and ζ_{min}, as it will be introduced in the numerical simulations section. Figure 2 show the control loop and the detail of the block diagram structure of the course keeping algorithm.

Theorem 2. *Consider the ship course dynamics described in (4) that complies with assumption 2. The application of the control law (21) to dynamic system (4) implies that the closed compact set Ω_e defined as*

$$\Omega_e = \{(e(t), \dot{e}(t)) \in \mathbb{R}^2 : |e(t)| < \frac{\mu}{|\cos(\theta)||\sin(\vartheta)|} \wedge |\dot{e}(t)| < \frac{\mu}{|\cos(\theta)||\cos(\vartheta)|}\} \tag{30}$$

is GUAS with μ, θ and ϑ given as

$$\theta = \text{atan}(\lambda) \tag{31}$$

$$\vartheta = \text{atan}(\frac{\lambda}{2}) \tag{32}$$

Proof. Application of control law (21) to dynamic system (4) creates the following cascade system.

$$\dot{e}(t) = -\lambda e(t) + s(t) \tag{33}$$

$$\dot{s}(t) = -\alpha s(t) - \gamma \bar{s}(t) + d(t) \tag{34}$$

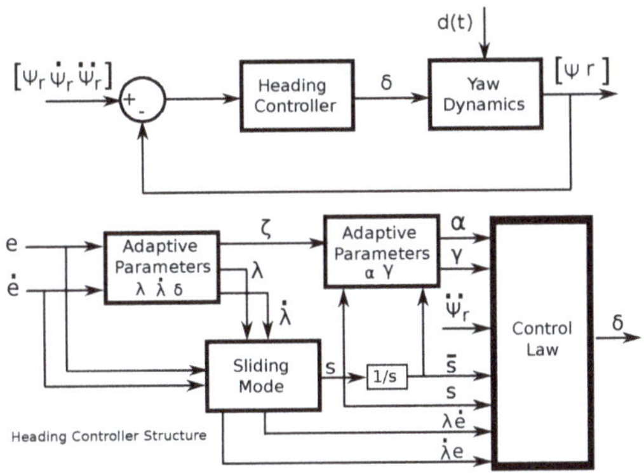

Figure 2. Course keeping control system and detail of heading control law block diagram.

The dynamics of $\dot{e}(t)$ when $s(t) = 0$ (dynamics of the yaw error at the sliding condition) are

$$\dot{e}(t) = -\lambda e(t) \tag{35}$$

with $\lambda > 0$. Therefore system (35) is GUAS, with exponential convergence.

Derivation of $\dot{s}(t)$ leads to

$$\ddot{s}(t) + \alpha \dot{s}(t) + \gamma s(t) + \dot{\alpha} s(t) + \dot{\gamma} \bar{s}(t) + \dot{d}(t) = 0$$

From (25), (28) and (23) it is obtained

$$\ddot{s}(t) + \alpha \dot{s}(t) + \frac{\alpha^2}{4} s(t) + \dot{\alpha} z(t) + \dot{d}(t) = 0$$

- **Case 1:** $|z| \leq \mu \wedge \alpha = \alpha_{min} \wedge \alpha = \alpha_{max}$
  Assuming the worst case scenario, that is, when $z > \mu \wedge \alpha = \alpha_{min}$, substitution of $\dot{\alpha} = 0$ implies that the second order dynamics related to $s(t)$ are

$$\ddot{s}(t) + \alpha_{min} \dot{s}(t) + \frac{\alpha_{min}^2}{4} s(t) + \dot{d}(t) = 0 \tag{36}$$

  Let's define a sliding vector state $\eta(t)$ as

$$\eta(t) = \begin{bmatrix} s(t) & \dot{s}(t) \end{bmatrix}^T \tag{37}$$

Dynamics of $\eta(t)$ are given by

$$\dot{\eta}(t) = A\eta(t) + F(t) \tag{38}$$

with

$$A = \begin{pmatrix} 0 & 1 \\ -\alpha_{min} & -\frac{\alpha_{min}^2}{4} \end{pmatrix} \tag{39}$$

and

$$F(t) = \begin{pmatrix} 0 \\ \dot{d}(t) \end{pmatrix} \tag{40}$$

Lets' define

$$p_1 = \frac{\alpha_{min}}{8} + \frac{2(\alpha_{min}+1)}{\alpha_{min}^2} \tag{41}$$

$$p_2 = \frac{0.5}{\alpha_{min}} \tag{42}$$

$$p_3 = \frac{2(\alpha_{min}+1)}{\alpha_{min}^3} \tag{43}$$

and P a symmetric positive definite matrix

$$P = \begin{pmatrix} p_1 & p_2 \\ p_2 & p_3 \end{pmatrix} \tag{44}$$

with determinant

$$|P| = \frac{\lambda^4 + 16\lambda^2 + 32\lambda + 16}{4\lambda^5} > 0 \tag{45}$$

It can be shown that

$$PA + A^T P = -Q \tag{46}$$

where Q is the identity matrix of size 2 × 2. Therefore, the selection of a Lyapunov candidate function

$$V(\eta) = \frac{1}{2}\eta^T P \eta \tag{47}$$

leads to

$$\dot{V}(\eta) = -\eta^T Q \eta + F^T P \eta \tag{48}$$

$$= -\eta^T Q \eta - \dot{d}(t)(p_2 s(t) + p_3 \dot{s}(t)) \tag{49}$$

Let's define

$$p^* = \sqrt{2}\max(p_2, p_3) \tag{50}$$

Therefore

$$|\dot{d}(t)(p_2 s(t) + p_3 \dot{s}(t))| \leq p^* \Delta ||\eta|| \tag{51}$$

Applying (51) and assumption 2 it is obtained

$$\dot{V}(\eta) < -\gamma_Q^{min} ||\eta||^2 + p^* \Delta ||\eta|| \tag{52}$$

Note that $\gamma_Q^{min} = 1$, thus the closed set Ω_η, which includes the origin, defined as

$$\Omega_\eta = \{\eta(t) \in \mathbb{R}^2 : ||\eta|| \leq p^*\Delta\} \tag{53}$$

is GUAS with exponential convergence (see [52]). The values of p^* and Δ determines the size of the stable closed set, so that this condition limits how the algorithm may be applied. Inside Ω_η there are two possible cases

- $\text{sign}(s\dot{s}) < 0$: implies that $|s(t)| \to 0$, that is, the dynamics is stable an converges to the origin, with a value of α adjusted to keep this condition.
- $\text{sign}(s\dot{s}) \geq 0$: implies that $|s(t)|$ grows inside Ω_η, that is, with an upper bound, or its value is stationary. According to (23) and (1), there exist an instant where the condition $\text{sign}(z)\,\text{sign}(s) > 0$ is met, which implies that α grows, that is, condition $\dot{\alpha} \neq 0$ is achieved. Therefore, α grows only when it is needed to keep the sliding mode condition at steady state, which is related to the performance given by the value of λ_{max}.

- **Case 2:** $|z| > \mu \wedge \alpha_{min} < \alpha < \alpha_{max}$
  Substitution of $\dot{\alpha} \neq 0$ from (24) implies that the second order dynamics equation related to $s(t)$ is

$$\ddot{s}(t) + \alpha\dot{s}(t) + \frac{\alpha^2}{4}s(t) + \kappa|z|^{\zeta+1}\,\text{sign}(s) + \dot{d}(t) = 0 \tag{54}$$

Applying assumption 2 and condition $|z| > \mu$ implies that

$$\kappa|z|^{\zeta+1}\,\text{sign}(s) + \dot{d} = \rho_z(t)s \tag{55}$$

with $\rho_z(t) > 0$. Accordingly, because of assumption 2, the characteristic polynomial of (54) is *Hurwitz* for all $z(t) \notin \Omega_z$ where

$$\Omega_z = \{z(t) \in \mathbb{R} : |z(t)| < \mu\} \tag{56}$$

with μ defined in (27). This implies that (34) is GUAS with respect to the closed set Ω_z. Note that dynamics in (54) can be viewed as a second order linear dynamics with adaptive critical damping (exponential convergence related to the fastest response with no overshooting), being perturbed by the overestimation $\rho_z s$ caused by the compensation of the unknown term. The roots of the perturbed solution of (54) are given by

$$s_{1,2}^* = \frac{\alpha}{2} \pm j\sqrt{\rho_z(t)} \tag{57}$$

A condition of the following form can be used to avoid chattering (high frequency oscillations caused by a large imaginary value in the pole position as a result of overestimation) at the steady-state response.

$$|\kappa|z|^{\zeta+1}\,\text{sign}(s) + \dot{d}(t)| \leq \kappa|z|^{\zeta+1} + \Delta < \rho_z^{max} \tag{58}$$

This provides an upper bound of the perturbation generated at the dynamics with respect to the solution with $\kappa = 0$ and $d(t) = 0$. In order to estimate the correlation between the sampling time τ and the natural frequency $\sqrt{\rho_z^{max}}$ (in $\frac{rad}{s}$), we must verify that the frequency given by the *Nyquist-Shannon* sampling theorem (the maximum operating frequency for a system with sampling time τ) does not create a change in sign in $s(t)$ at the limit condition $|z(t)| = \mu$, that is

$$\sqrt{\rho_z^{max}} \leq \frac{\pi}{\tau}\mu \tag{59}$$

Applying condition (58) an upper bound for κ as a function of Δ, τ and the absolute value of $z(t)$ is obtained as

$$\kappa \leq \frac{(\frac{\pi\mu}{\tau})^2 - \Delta}{|z|^{\zeta+1}} \tag{60}$$

This constraint provides a limit on the application of the method that employs the bound Δ of the disturbance derivative, the sampling time τ and, taking into consideration relation (27), the required precision μ.
Inside Ω_z we have that

$$|s(t) + \frac{\alpha}{2}\bar{s}(t)| < \mu$$

which geometrically entails:

$$|\bar{s}(t)| < \frac{\mu}{|\sin(\vartheta)|} \tag{61}$$

$$|s(t)| < \frac{\mu}{|\cos(\vartheta)|} \tag{62}$$

with ϑ defined in (32). Inside Ω_s we have that

$$|\dot{e}(t) + \lambda e(t)| < \frac{\mu}{|\cos(\vartheta)|}$$

Following the previous approach implies that:

$$|e(t)| < \frac{\mu}{|\cos(\vartheta)||\sin(\theta)|} \tag{63}$$

$$|\dot{e}(t)| < \frac{\mu}{|\cos(\vartheta)||\cos(\theta)|} \tag{64}$$

with θ defined in (31).

Applying Theorem 1 with

$$\begin{aligned}
\phi_1(||e||) &= k_1 e^2 \\
\phi_2(||e||) &= k_2 e^2 \\
\phi_3(||e||) &= k_3 |e| \\
W(e) &= k_4 e^2 \\
V(e) &= \frac{1}{2} e^2 \\
\zeta(e) &= \frac{k_5}{k_4 |e|}
\end{aligned}$$

where $k_1 < 0.5$, $k_2 > 0.5$, $k_3 > 1.0$, $k_4 < \lambda$ and $k_5 > 1.0$, entails that cascade system given in (33) and (34) is GUAS with respect to the closed compact sets Ω_e and Ω_s, respectively. □

4. Numerical Simulations

In this section we introduce numerical simulations of the course keeping problem with parameters given in Table 1 and being executed under the following assumption.

Assumption 4. *The numerical simulations are executed using the explicit Euler method with fixed sampling time* $\tau = 0.1$ *s.*

Table 1. Model parameters.

Parameter	Value
K	0.21
T	107.76
a_1	13.17
a_2	16,323.46

4.1. Constant Yaw Reference

This test is presented in [12] with a required a change in the yaw orientation angle from zero initial condition up to 50 degrees assuming that $d(t) = 0$. Table 2 show the parameters used in [12].

Table 2. Nonlinear concise backstepping controller parameters.

Parameter	Value
k_1	0.0017
ω	0.6000

Based on this results the parameter a_2 of the synergetic controller presented in [10] is changed to achieve the same MIA at the end of the simulation. Table 3 show the parameters used with this algorithm.

Table 3. Synergetic controller parameters.

Parameter	Value
a_1	0.090
a_2	1.891
T_1	28.000

The parameters of the AISM algorithm are obtained as follows

- Consider a settling time $t_s = 150s$, a maximum desired yaw rate $r_{max} = \frac{0.70\pi}{180}$ degrees per second and a required precision $\mu = 1.0 \times 10^{-6}$.
- The value of $\alpha(0)$ is obtained assuming an exponential convergence of the error from initial condition $e(0)$ to desired precision μ with a desired settling time t_s

$$\alpha(0) = -1.25\frac{log(\frac{\mu}{|e(0)|})}{t_s} = 0.0756 \tag{65}$$

The values of α_{min} and α_{max} are selected as

$$\alpha_{min} = \alpha(0) \tag{66}$$

$$\alpha_{max} = 5(\alpha_{min} + \Delta) \tag{67}$$

- The value of λ_{min} is related to the initial conditions of the problem and the maximum desired yaw rate as

$$\lambda_{min} = \frac{r_{max}}{|e(0)|} = 0.014 \tag{68}$$

and λ_{max} is calculated as

$$\lambda_{max} = 2.0\lambda_{min} = 0.028 \tag{69}$$

- The value of κ must be higher than Δ in order to obtain a small value for μ. Due ti the low-pass filtering properties of (54), the value of Δ can be further refined by estimating the cut-off frequency $\omega_c(t)$ of the second order system related to $s(t)$

$$\omega_c(t) = \frac{\alpha(t)}{2} \tag{70}$$

Therefore κ is calculated as an adaptive gain that takes account of ω_c and the desired precision

$$\kappa(t) = \frac{\alpha(t)}{2\mu^{\frac{1}{\zeta+1}}} \tag{71}$$

- Simulations are used to set the values of ζ_{min} and ζ_{max} such that the value of the performance index MIA is equal to the value achieved with benchmark chosen controllers at the conclusion of the test period.

$$\zeta_{min} = 0.800$$
$$\zeta_{max} = 1.685$$

This condition generates an adequate adaption of the value of ζ that allows to obtain the desired power factor profile with respect to the absolute value of $e(t)$.

Figure 3 shows the states and control effort, and it can be seen that all of the solutions have a comparable setting time. The yaw error evolves similarly in all circumstances, however AISM can achieve the same high accuracy in steady state than Synergetic control with less control effort and a lower maximum yaw rate.

Figure 4 depicts the progression of performance indices over time, with a detailed view of the MIA performance index at the end of the test and final numerical values in Table 4. The evolution of the adaptive parameters employed in the proposed AISM method is detailed in Figure 5. In the case of $d(t) = 0$, the value of the performance parameter λ is adjusted to match the intended low/high gain profile, while the value of gain α grows until it reaches the sliding condition, then falls to its lower bound.

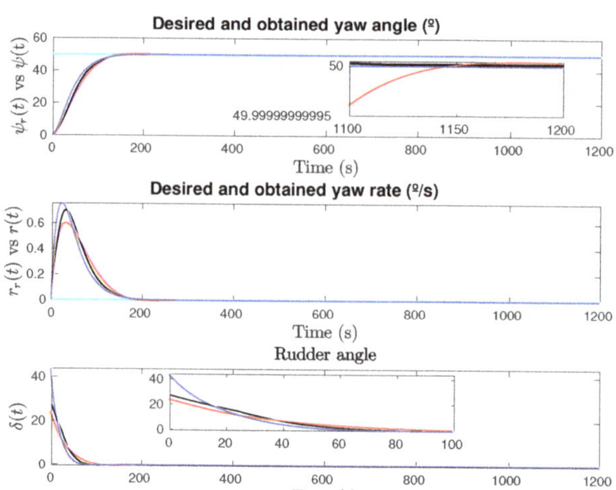

Figure 3. Constant yaw reference test with $d(t) = 0$. States and control. Cyan line: Reference; Red line: Concise backstepping (Zhang et al.); Blue line: Synergetic (Muhammad et al.); Black line: Adaptive sliding mode (González-Prieto et al.).

Table 4. Constant yaw reference test with $d(t) = 0$. Performance indices.

Algorithm	MAE	MIA	MTV
Concise Backstepping [12]	0.042227	0.011348	3.6396×10^{-5}
Synergetic [10]	0.035641	0.011348	6.4436×10^{-5}
AISM	0.038468	0.011348	4.0722×10^{-5}

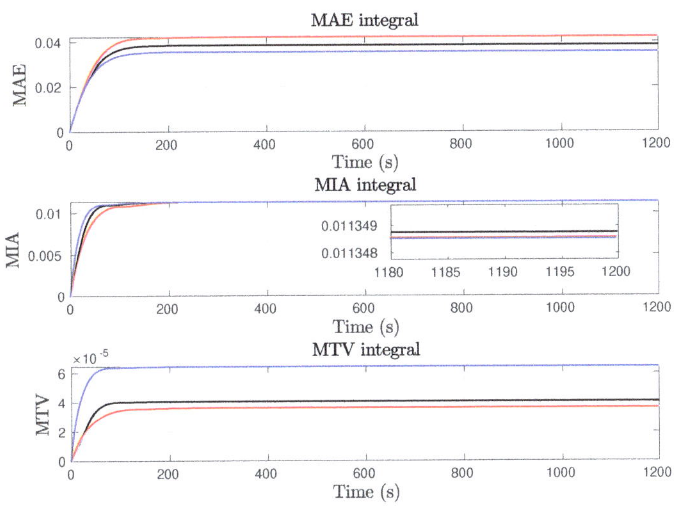

Figure 4. Constant yaw reference test with $d(t) = 0$. Performance indices evolution. Cyan line: Reference; Red line: Concise backstepping (Zhang et al.); Blue line: Synergetic (Muhammad et al.); Black line: Adaptive sliding mode (González-Prieto et al.).

Next, in order to test the robustness of the algorithms, the following disturbance is considered in (4)

$$
\begin{aligned}
d(t) = D[&\cos(\omega_d t) + 0.83\sin(3.29\omega_d t - 0.14) \\
&+ 1.23\cos(8.12\omega_d t + 0.26) \\
&+ 0.65\sin(1.37\omega_d t + 0.36)e^{\cos(2.21\omega_d t + 0.13)}]
\end{aligned} \tag{72}
$$

with

$$
\begin{aligned}
D &= 0.0025 \\
\omega_d &= 0.0703\frac{rad}{s}
\end{aligned}
$$

Figure 6 shows the states and control effort, and it is obvious that the suggested AISM negates the influence of the external disturbance, maintaining the intended performance at steady-state, and creating a rudder angle control that enables quick reaction attenuation without causing overshooting. The evolution of the sliding variable $s(t)$ and the external disturbance $d(t)$, introduced to evaluate the resilience qualities of the compared algorithms, is depicted in Figure 7.

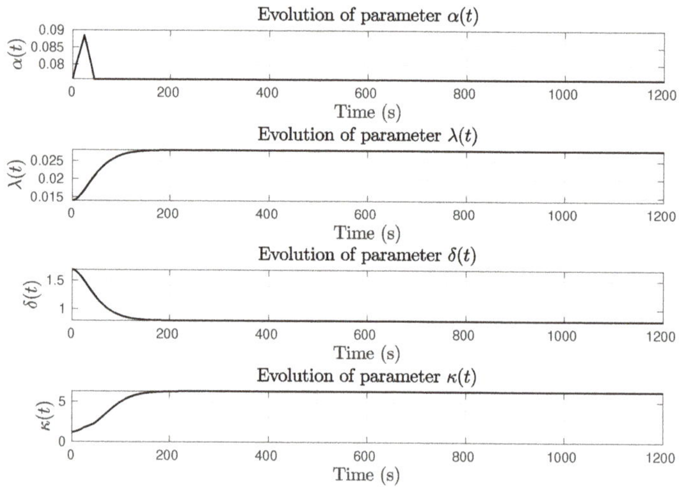

Figure 5. Constant yaw reference test with $d(t) = 0$. Adaptive parameters evolution.

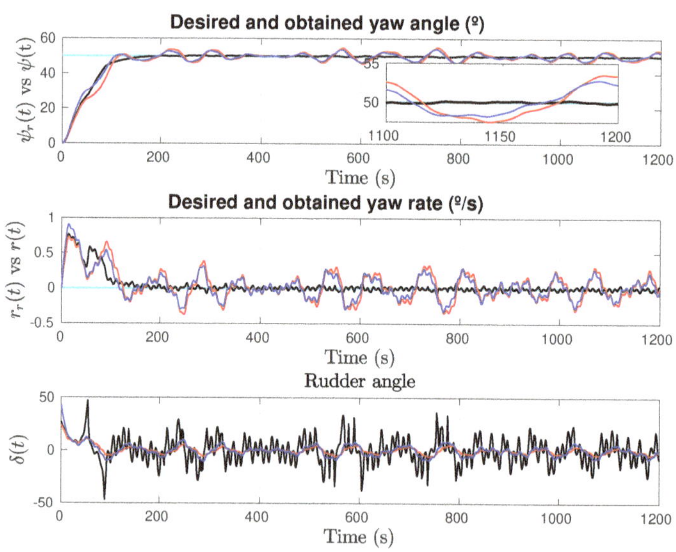

Figure 6. Constant yaw reference test with $d(t) \neq 0$. States and control. Cyan line: Reference; Red line: Concise backstepping (Zhang et al.); Blue line: Synergetic (Muhammad et al.); Black line: Adaptive sliding mode (González-Prieto et al.).

The evolution of the adaptive gains is seen in Figure 8. It is evident that the influence of the unknown disturbance is passed to α and κ in order to maintain steady-state performance, but it has no effect on lambda or the adaptive power factor delta. As a result, the reaction achieves a quick response to external disturbances in order to maintain a constant target performance related to the sliding condition $s(t) \approx 0$.

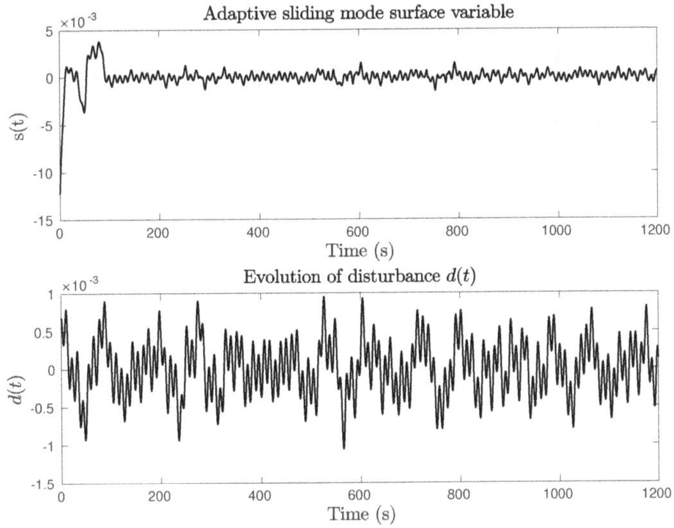

Figure 7. Constant yaw reference test with $d(t) \neq 0$. Sliding mode variable $s(t)$ and external disturbance $d(t)$.

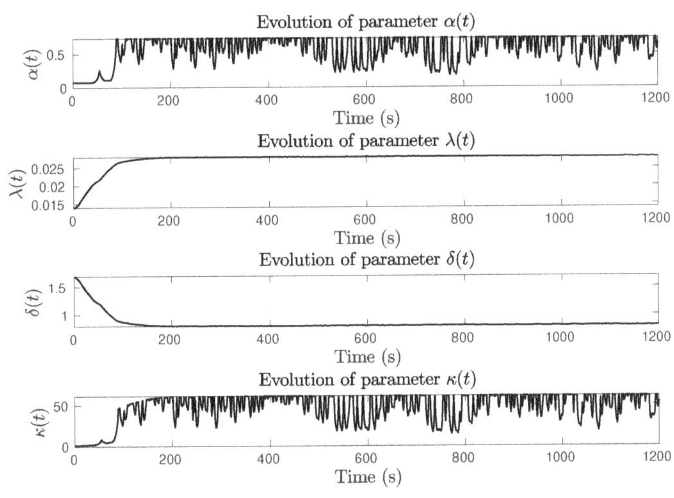

Figure 8. Constant yaw reference test with $d(t) \neq 0$. Adaptive parameters evolution.

4.2. Sinusoidal Yaw Reference

In this case, as in [12], the yaw reference to follow is a sinusoidal signal defined as

$$\omega_d = \frac{50\pi}{180.0}\sin\left(\frac{2\pi}{600.0}\right) \tag{73}$$

where the initial yaw angle is

$$\psi(0) = \frac{10\pi}{180.0} \tag{74}$$

States and control effort are provided in Figure 9 where it is clear that AISM is capable to follow the yaw reference with no appreciable delay keeping the desired settling time, and Figure 10 depicts the adaptive parameter's change over time in the unperturbed case with sinusoidal reference.

As in the previous test, results with sinusoidal reference are tested introducing disturbance (72). Figure 11 shows the states and control effort obtained in this case, where, as in the constant reference test, the steady-state performance and the settling time obtained with AISM are preserved despite the presence of the external unknown disturbance.

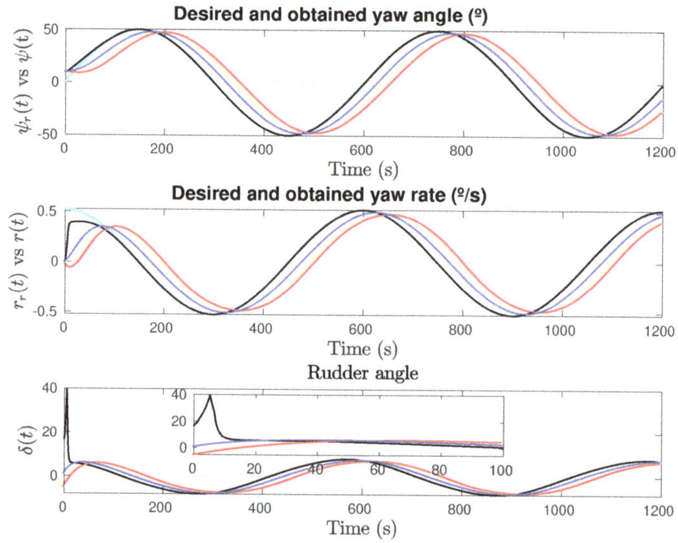

Figure 9. Sinusoidal yaw reference test with $d(t) = 0$. States and control. Cyan line: Reference; Red line: Concise backstepping (Zhang et al.); Blue line: Synergetic (Muhammad et al.); Black line: Adaptive sliding mode (González-Prieto et al.).

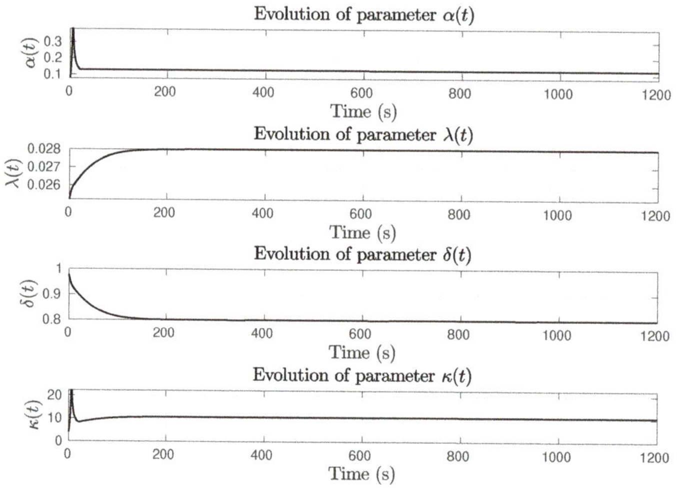

Figure 10. Sinusoidal yaw reference test with $d(t) = 0$. Adaptive parameters evolution.

Figure 12 shows the evolution of the sliding variable and the external disturbance input $d(t)$. The time evolution of adaptive parameters in Figure 13 mirrors the behavior in the case of constant yaw reference in case of $d(t) = 0$. This characteristic, like in constant yaw reference scenario, generates quick reactions to external disturbances, maintaining a desired fixed performance and avoiding a delayed response.

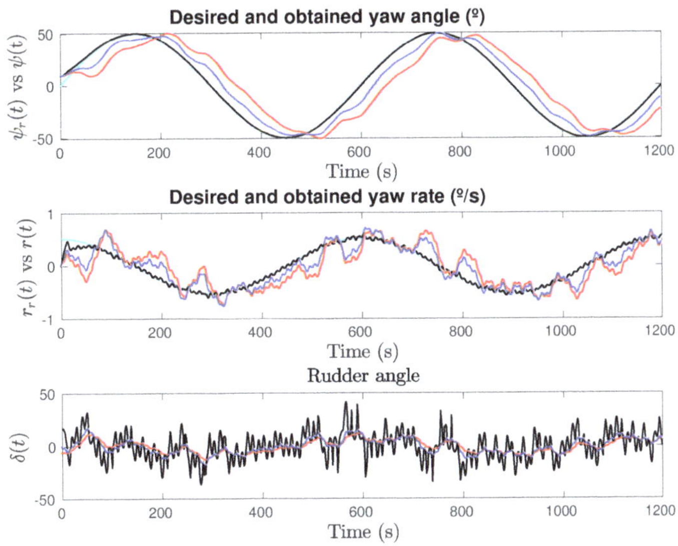

Figure 11. Sinusoidal yaw reference test with $d(t) \neq 0$. States and control. Cyan line: Reference; Red line: Concise backstepping (Zhang et al.); Blue line: Synergetic (Muhammad et al.); Black line: Adaptive sliding mode (González-Prieto et al.).

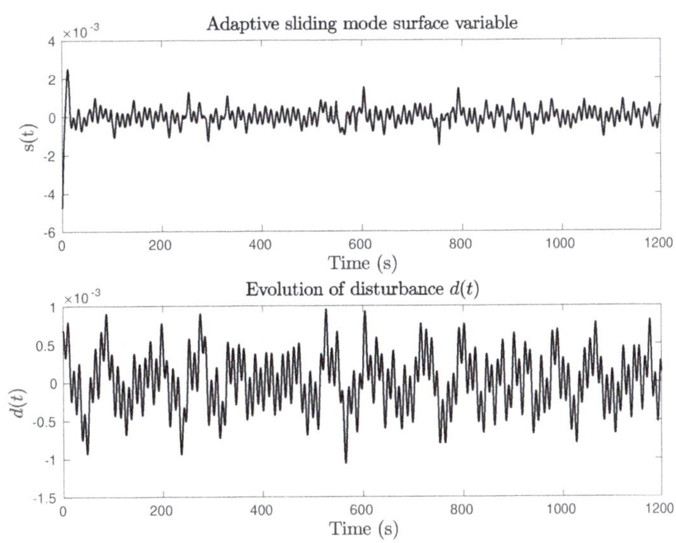

Figure 12. Sinusoidal yaw reference test with $d(t) \neq 0$. Sliding mode variable $s(t)$ and external disturbance $d(t)$.

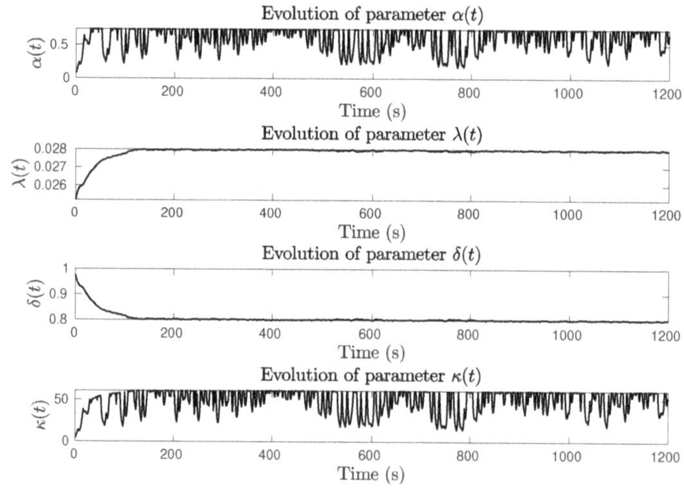

Figure 13. Sinusoidal yaw reference test with $d(t) \neq 0$. Adaptive parameters evolution.

5. Conclusions

We present an approach to developing an adaptive integral sliding mode procedure to design a nonlinear controller for the course keeping of surface vehicles. A solution has been proposed that is based on the application of adaptive gains that change the sliding surface's dumping properties, resulting in a low/high gain profile so that it can overcome the need for large control inputs at initial conditions while achieving a higher gain at steady state.

Results obtained in numerical simulations demonstrate that the proposed AISM algorithm achieves the desired performance with fixed and time-varying references that cancel out external disturbances. A fixed parameter configuration is used to evaluate the performance based on the settling time, maximum allowable yaw rate, and steady state precision. Due to its robustness, the algorithm achieves the desired response without requiring the development of a new parameter configuration for each type of test.

An advantage of the method is its robustness with respect to an overestimation of Δ: the performance is not highly degraded if this bound is not accurately known. However, choosing an overly large value might cause oscillations in the response of the estimation error.

In order to integrate an optimal point of view in the design of the adaptive parameters, a deepest study of the function that can determine the adaptive values of $\lambda(e)$ and $\zeta(e)$ is an interesting open problem that can be discussed from the standpoint of model predictive control.

The extension of this procedure with the assumption of partial state feedback will be addressed in future researches by means of the application of an adaptive integral sliding mode observers.

Author Contributions: J.A.G.-P.: Conceptualization, Methodology, Software, Formal analysis, Writing—original draft, C.P.-C.: Formal analysis, Writing—original draft, Y.S.: Formal analysis, Writing—original draft. All authors have read and agreed to the published version of the manuscript.

Funding: J. A. González-Prieto and C. Pérez-Collazo acknowledges funding from the Defense University Center at the Spanish Naval Academy, Spanish Ministry of Defense, TRABODIT project (PICUD-2021-01).

Institutional Review Board Statement: Not applicable.

Informed Consent Statement: Not applicable.

Data Availability Statement: Not applicable.

Conflicts of Interest: The authors declare that they have no known competing financial interests or personal relationships that could have appeared to influence the work reported in this paper.

Abbreviations

The following abbreviations are used in this manuscript:

USV	Unmanned Surface Vehicle
IMO	International Maritime Organisation
LOS	Line of Sight
SMC	Sliding Mode Control
AISM	Adaptive Integral Sliding Mode
GUAS	globally uniformly asymptotically stable
MAE	Mean Absolute Error
MIA	Mean Integral Absolute
MTV	Mean Total Variation

References

1. Fossen, T.I. *Handbook of Marine Craft Hydrodynamics and Motion Control*; John Wiley & Sons: Hoboken, NJ, USA, 2011.
2. Do, K.D.; Pan, J. *Control of Ships and Underwater Vehicles: Design for Underactuated and Nonlinear Marine Systems*; Springer Science & Business Media: Berlin/Heidelberg, Germany, 2009.
3. Liu, Z. Ship adaptive course keeping control with nonlinear disturbance observer. *IEEE Access* **2017**, *5*, 17567–17575. [CrossRef]
4. Ejaz, M.; Chen, M. Sliding mode control design of a ship steering autopilot with input saturation. *Int. J. Adv. Robot. Syst.* **2017**, *14*, 1729881417703568. [CrossRef]
5. Wan, L.; Su, Y.; Zhang, H.; Tang, Y.; Shi, B. Global fast terminal sliding mode control based on radial basis function neural network for course keeping of unmanned surface vehicle. *Int. J. Adv. Robot. Syst.* **2019**, *16*, 1729881419829961. [CrossRef]
6. Rodriguez, J.; Castañeda, H.; Gordillo, J.L. Design of an adaptive sliding mode control for a micro-AUV subject to water currents and parametric uncertainties. *J. Mar. Sci. Eng.* **2019**, *7*, 445. [CrossRef]
7. Khaled, N.; Chalhoub, N.G. A self-tuning guidance and control system for marine surface vessels. *Nonlinear Dyn.* **2013**, *73*, 897–906. [CrossRef]
8. Budak, G.; Beji, S. Controlled course-keeping simulations of a ship under external disturbances. *Ocean. Eng.* **2020**, *218*, 108126. [CrossRef]
9. Zhang, X.-K.; Zhang, Q.; Ren, H.-X.; Yang, G.-P. Linear reduction of backstepping algorithm based on nonlinear decoration for ship course-keeping control system. *Ocean. Eng.* **2018**, *147*, 1–8. [CrossRef]
10. Islam, M.M.; Siffat, S.A.; Ahmad, I.; Liaquat, M. Robust integral backstepping and terminal synergetic control of course keeping for ships. *Ocean. Eng.* **2021**, *221*, 108532. [CrossRef]
11. Zhang, X.-K.; Han, X.; Guan, W.; Zhang, G.-Q. Improvement of integrator backstepping control for ships with concise robust control and nonlinear decoration. *Ocean. Eng.* **2019**, *189*, 106349. [CrossRef]
12. Zhang, Q.; Zhang, X. Nonlinear improved concise backstepping control of course keeping for ships. *IEEE Access* **2019**, *7*, 19258–19265. [CrossRef]
13. Zheng, Z.; Sun, L. Path following control for marine surface vessel with uncertainties and input saturation. *Neurocomputing* **2016**, *177*, 158–167. [CrossRef]
14. Zhang, X.-K.; Zhang, G.-Q. Design of ship course-keeping autopilot using a sine function-based nonlinear feedback technique. *J. Navig.* **2016**, *69*, 246–256. [CrossRef]
15. Zhao, H.; Zhang, X.; Han, X. Nonlinear control algorithms for efficiency-improved course keeping of large tankers under heavy sea state conditions. *Ocean. Eng.* **2019**, *189*, 106371. [CrossRef]
16. Zhang, Q.; Zhang, X.-K.; Im, N.-K. Ship nonlinear-feedback course keeping algorithm based on MMG model driven by bipolar sigmoid function for berthing. *Int. J. Nav. Archit. Ocean. Eng.* **2017**, *9*, 525–536. [CrossRef]
17. Zhang, Q.; Zhang, X.-K.; Im, N.-K. Adaptive neural path-following control for underactuated ships in fields of marine practice. *Ocean. Eng.* **2015**, *104*, 558–567. [CrossRef]
18. Zhang, J.; Sun, T.; Liu, Z. Robust model predictive control for path-following of underactuated surface vessels with roll constraints. *Ocean. Eng.* **2017**, *143*, 125–132. [CrossRef]
19. Sharma, S.K.; Sutton, R.; Motwani, A.; Annamalai, A. Non-linear control algorithms for an unmanned surface vehicle. *Proc. Inst. Mech. Eng. Part M J. Eng. Marit. Environ.* **2014**, *228*, 146–155. [CrossRef]
20. Zhang, G.; Li, J.; Jin, X.; Liu, C. Robust Adaptive Neural Control for Wing-Sail-Assisted Vehicle via the Multiport Event-Triggered Approach. *IEEE Trans. Cybern.* **2021**, 1–13. 10.1109/TCYB.2021.3091580. [CrossRef]
21. Yan, Z.; Zhang, X.; Zhu, H.; Li, Z. Course-keeping control for ships with nonlinear feedback and zero-order holder component. *Ocean. Eng.* **2020**, *209*, 107461. [CrossRef]
22. Utkin, V.; Guldner, J.; Shi, J. *Sliding Mode Control in Electro-Mechanical Systems*; CRC Press: Boca Raton, FL, USA, 2009.
23. Edwards, C.; Spurgeon, S. *Sliding Mode Control: Theory And Applications*; CRC Press: Boca Raton, FL, USA, 1998.

24. Levant, A. Robust exact differentiation via sliding mode technique. *Automatica* **1998**, *34*, 379–384. [CrossRef]
25. Edwards, C.; Spurgeon, S.K.; Patton, R.J. Sliding mode observers for fault detection and isolation. *Automatica* **2000**, *36*, 541–553. [CrossRef]
26. Feng, Y.; Xinghuo, Y.; Zhihong, M. Non-singular terminal sliding mode control of rigid manipulators. *Automatica* **2002**, *38*, 2159–2167. [CrossRef]
27. Nersesov, S.G.; Ashrafiuon, H.; Ghorbanian, P. On estimation of the domain of attraction for sliding mode control of underactuated nonlinear systems. *Int. J. Robust Nonlinear Control* **2014**, *24*, 811–824. [CrossRef]
28. Hao, Y.; Yi, J.; Zhao, D.; Qian, D. Robust control using incremental sliding mode for underactuated systems with mismatched uncertainties. In Proceedings of the American Control Conference, Seattle, WA, USA, 11–13 June 2008; pp. 532–537.
29. Liu, J.; Laghrouche, S.; Harmouche, M.; Wack, M. Adaptive-gain second-order sliding mode observer design for switching power converters. *Control Eng. Pract.* **2014**, *30*, 124–131. [CrossRef]
30. Yang, J.; Li, S.; Yu, X. Sliding-mode control for systems with mismatched uncertainties via a disturbance observer. *IEEE Trans. Ind. Electron.* **2013**, *60*, 160–169. [CrossRef]
31. Oliveira, T.R.; Cunha, J.P.V.; Hsu, L. Adaptive sliding mode control for disturbances with unknown bounds. In Proceedings of the 14th International Workshop on Variable Structure Systems (VSS), Nanjing, China, 1–4 June 2016; pp. 59–64.
32. Hsu, L.; Oliveira, T.R.; Cunha, J.P.V.; Yan, L. Adaptive unit vector control of multivariable systems using monitoring functions. *Int. J. Robust Nonlinear Control* **2019**, *29*, 583–600. [CrossRef]
33. Chen, M.; Wu, Q.; Cui, R. Terminal sliding mode tracking control for a class of SISO uncertain nonlinear systems. *ISA Trans.* **2013**, *52*, 198–206. [CrossRef]
34. Wang, W.; Liu, X.D.; Yi, J.Q. Structure design of two types of sliding-mode controllers for a class of under-actuated mechanical systems. *IET Control Theory Appl.* **2007**, *1*, 163–172. [CrossRef]
35. Thanh, H.L.N.N.; Mung, N.X.; Nguyen, N.P.; Phuong, N.T. Perturbation observer-based robust control using a multiple sliding surfaces for nonlinear systems with influences of matched and unmatched uncertainties. *Mathematics* **2020**, *8*, 1371. [CrossRef]
36. Alattas, K.A.; Mobayen, S.; Din, S.U.; Asad, J.H.; Fekih, A.; Assawinchaichote, W.; Vu, M.T. Design of a non-singular adaptive integral-type finite time tracking control for nonlinear systems with external disturbances. *IEEE Access* **2021**, *9*, 102091–102103. [CrossRef]
37. Li, Y.-X.; Yang, G.-H. Adaptive integral sliding mode control fault tolerant control for a class of uncertain nonlinear systems. *IET Control Theory Appl.* **2018**, *12*, 1864–1872. [CrossRef]
38. Liu, D.; Yang, G.-H. Prescribed performance model-free adaptive integral sliding mode control for discrete-time nonlinear systems. *IEEE Trans. Neural Netw. Learn. Syst.* **2018**, *30*, 2222–2230. [CrossRef]
39. Edwards, C.; Shtessel, Y.B. Adaptive continuous higher order sliding mode control. *Automatica* **2016**, *65*, 183–190. [CrossRef]
40. Feng, Y.; Han, F.; Yu, X. Chattering free full-order sliding-mode control. *Automatica* **2014**, *50*, 1310–1314. [CrossRef]
41. Bandyopadhyay, B.; Deepak, F.; Kim, K. *Sliding Mode Control Using Novel Sliding Surfaces*; Springer: Berlin/Heidelberg, Germany, 2009; Volume 392.
42. González, J.A.; Barreiro, A.; Dormido, S.; Banos, A. Nonlinear adaptive sliding mode control with fast non-overshooting responses and chattering avoidance. *J. Frankl. Inst.* **2017**, *354*, 2788–2815. [CrossRef]
43. González, J.A.; Barreiro, A.; Dormido, S. A practical approach to adaptive sliding mode control. *Int. J. Control* **2019**, *17*, 2452–2461. [CrossRef]
44. Polyakov, A. *Generalized Homogeneity in Systems and Control*; Springer: Berlin/Heidelberg, Germany, 2020.
45. Mojallizadeh, M.R.; Brogliato, B.; Acary, V. Time-discretizations of differentiators: Design of implicit algorithms and comparative analysis. *Int. J. Robust Nonlinear Control* **2021**, *31*, 7679–7723. [CrossRef]
46. Witkowska, A.; Śmierzchalski, R. Designing a ship course controller by applying the adaptive backstepping method. *Int. J. Appl. Math. Comput. Sci.* **2012**, *22*, 985–997. [CrossRef]
47. Singh, Y.; Sharma, S.; Sutton, R.; Hatton, D. Path Planning of an Autonomous Surface Vehicle Based on Artificial Potential Fields in a Real Time Marine Environment; Cardiff University Press: Cardiff, UK, 2017.
48. Polvara, R.; Sharma, S.; Wan, J.; Manning, A.; Sutton, R. Obstacle avoidance approaches for autonomous navigation of unmanned surface vehicles. *J. Navig.* **2018**, *71*, 241–256. [CrossRef]
49. Singh, Y.; Sharma, S.; Sutton, R.; Hatton, D. Optimal path planning of an unmanned surface vehicle in a real-time marine environment using a dijkstra algorithm. In *Marine Navigation*; CRC Press: Boca Raton, FL, USA, 2017; pp. 241–256. 399–402 .
50. Singh, Y.; Sharma, S.; Sutton, R.; Hatton, D.; Khan, A. A constrained A* approach towards optimal path planning for an unmanned surface vehicle in a maritime environment containing dynamic obstacles and ocean currents. *Ocean. Eng.* **2018**, *169*, 187–201. [CrossRef]
51. Panteley, E.; Loria, A. On global uniform asymptotic stability of nonlinear time-varying systems in cascade. *Syst. Control Lett.* **1998**, *33*, 131–138. [CrossRef]
52. Khalil, H.K.; Grizzle, J.W. *Nonlinear Systems*; Prentice Hall: Hoboken, NJ, USA, 1996.

Journal of *Marine Science and Engineering*

Article

Cooperation between Sea Ports and Carriers in the Logistics Chain

Elena A. Lezhnina * and Yulia E. Balykina *

Faculty of Applied Mathematics and Control Processes, Saint Petersburg State University, 199034 Saint Petersburg, Russia
* Correspondence: e.lezhnina@spbu.ru (E.A.L.); j.balykina@spbu.ru (Y.E.B.)

Abstract: In a modern economy, international trade is an important factor in the development of various regions. Shipping is one of the most important elements of the global supply chain. However, after the economic crisis of 2008, global shipping revenues plummeted. One way to restore profitability is the consolidation of shipping routes and the globalization of shipping lines. As container transport lines move to larger ships, the structure of the delivery route becomes a structure with intermediate points. This trend put forward higher demands on the port infrastructure, which aggravated the competition between regional ports, as well as ports that could degrade into a large cargo consolidation port. The economic advantage is enhanced by cooperation between shipping lines and ports. Thus, ports and shipping lines in the same supply chain can be mutually beneficial partners. The study analyses the effectiveness of horizontal and vertical cooperation between ports and carriers. As a source of information, a review of the literature on this issue, expert opinions, and statistical data is taken. Next, a mathematical model is built on the basis of cooperative game theory, and numerical analysis is carried out. The results show that the strategy of cooperation of shipping lines strongly depends on the situation with the supply and demand of vessels. A port that interacts with shipping lines will significantly reduce port charges, which creates the advantage of receiving more port requests. However, cooperation may lead to losses for the port, so a redistribution of profits is necessary to maintain the coalition.

Keywords: cooperative game theory; supply chain management; supply disruption

Citation: Lezhnina, E.A.; Balykina, Y.E. Cooperation between Sea Ports and Carriers in the Logistics Chain. *J. Mar. Sci. Eng.* **2021**, *9*, 774. https://doi.org/10.3390/jmse9070774

Academic Editor: Mihalis Golias

Received: 17 June 2021
Accepted: 14 July 2021
Published: 16 July 2021

1. Introduction

With the development of the world economy, international trade has become an important factor in the economic development of different regions. As a key element of the global trade chain, the shipping industry covers more than 80 percent of the world trade. However, after the global financial crisis of 2008, the shipping industry faced problems of overcapacity. According to various data, by 2008 about 35–40% of container capacity was already excessive [1–3]. To restore profitability, it has become common practice to consolidate sea routes and globalize shipping lanes. For example, [3] discusses the cooperation of European seaports: the advantages, limitations, and development prospects. In [4], the activities of the largest alliances of sea carriers are analyzed, and the profitability of creating such alliances is shown.

Shipping companies have taken steps to unite their fleet in a coalition. The coalition of carriers can receive lower prices for the port service, and use the vessels more economically, fully loading them, rather than using partially loaded ones. In [5], the game theory approach is applied, showing the profitability of the sea carriers' merger, the problems of the stability of alliances, and the distribution of the total profit between the participants. A significant advantage of cooperation between participants in the supply chain is the ability to reduce losses in case of accidental interruptions in supplies, which can occur for various reasons (e.g., a fully loaded port cannot accept cargo for processing, equipment failure may occur, weather conditions are unfavorable, etc.). In [3], a theoretical model of the supply chain is proposed, and the role of the port as a participant in this chain is discussed. In [6], the efficiency of supply chain management is discussed. In [7,8], models are considered which show the importance of coordinating the work of all participants in chain.

Consolidation of shipping lanes led to significant changes in the shipping industry. Cooperation agreements between ports and carriers have led to changes in the role of the ports. Larger ports stood out as hubs dealing with the largest ships. Smaller ports began to serve hub ports, "collecting" cargo from land transport lines. In modern conditions, peripheral ports lose in competition with hub ports. However, the cooperation of sea routes and ports, uniting peripheral ports and hubs in a single coalition, gives ports an opportunity to increase the level of loading and unloading services, and allows sea routes to receive a stable income from the port's operation. It also protects peripheral ports from closure, saving jobs.

The goal of this study is to analyze the possibilities of various participant cooperation in the logistics chain in order to increase profits and reduce costs. This article discusses a logistic chain that includes a hub port, peripheral ports, and land carriers. It is shown that the cooperative work of the participants in the chain brings significant savings in logistics costs.

2. Theoretical Background

Global changes in the carrier market are of great interest. In [3], a theoretical model of the supply chain is proposed and the role of the port as a participant in this chain is discussed. Effective supply chain management methods are discussed in [6]. Paying close attention to this topic helps to avoid the problems of shortages or surpluses of goods, as well as mitigate problems in case of possible supply disruptions.

The complete shipping chain in the general case is as follows (Figure 1):

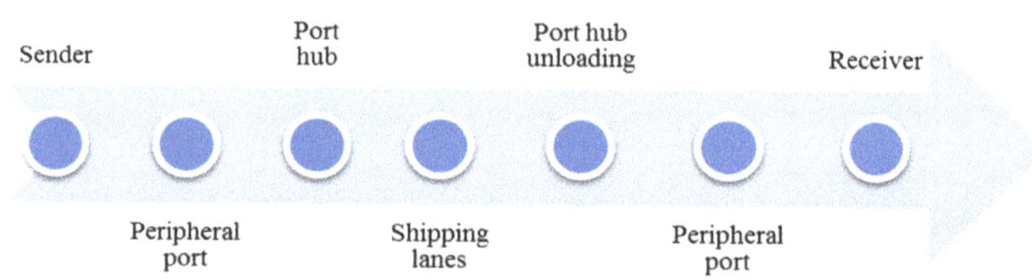

Figure 1. The scheme of shipping supply chain.

At each stage, horizontal cooperation of several participants is possible. One might also consider vertical cooperation, which brings together participants from different parts of the logistics chain.

2.1. Customer Cooperation at the Stage of Placing an Order

Cost reduction in this type of cooperation is achieved through the formation of bigger quantities of goods, which allows the use of larger vessels and, as a result, reduces the cost of delivery. Water shipment is more profitable than land transportation, provided that the consignment of goods is large enough. Therefore, ports often act as cargo aggregators, collecting several orders for a large vessel. In addition, at the stage of consignment formation, one can use the optimization model of the delivery scheme. Individual customers cannot always form batches that are optimal in terms of the economic order quantity (EOQ), as they are bound by standard container sizes. Joint orders allow more flexibility to determine the frequency and size of orders. In detail the task of forming the optimal size of the order is considered in [9].

2.2. Port Cooperation and Competition

Let us assume that the economical region under consideration has a main port that handles large vessels. In addition, there are several peripheral ports. When creating an order, a consumer may be guided by various reasons when choosing a port: contracts, service convenience, special offers, etc. In addition, ports have different capabilities for processing orders. Queues and delays are possible for this reason. The benefits of combining are due to more coordinated cargo handling.

Such cooperation is a method used by market players in an attempt to reduce a high level of competitiveness [10]. Horizontal integration between ports can lead to lower freight rates and more efficient logistic control, thus increasing the demand for port services. Some authors suggest that cooperation between ports can have a positive impact on the overall competitiveness of ports in the region, leading to overall market growth in the region.

Ports compete with one another for customers. In [10], the authors discuss the advantages of cooperation over competition. The main indicators of the port functioning are the number of sites for unloading, the speed of cargo handling, and the cost of service. Competition among the ports leads to the situation in which the main flow of goods in the region goes through the "winning" port. To survive, small ports switch to peripheral shipments. At the same time, it is beneficial to enter into a peripheral port–port hub coalition. The coalition is beneficial to both parties, as both types of ports are loaded with work.

Port competition is more complex than it seems to be. Not all ports have the necessary capacity to ensure the loading and unloading of modern large vessels. At the same time, already allocated hubs that have intercepted the main flow of goods make competition very difficult, leaving the rest of competitors few opportunities for development.

What are the options for alliances around the hub? The first option is a contract between a peripheral port and a hub port. As a result, the hub receives the flow of goods, while the peripheral port gains a possible reduction in tariffs for service and priority service. However, if the port hub has limited capacity to handle cargo, holdups may occur, and the competition between the ports will escalate. The cause of such congestion may be an insufficient number of unloading sites, especially for large vessels, slow service, etc. In this case, the ports compete with one another. Customers are attracted by higher levels of service or lower prices. Price competition in this case is well described in terms of Bertrand oligopoly.

The second type of possible cooperation between the ports is the creation of coalitions of equal ports. This can lead to better integration in the supply chain, lower costs, and the elimination of redundant links, as well as more flexible route planning conditions.

In addition to reducing costs, port cooperation helps reduce the impact of possible supply disruptions. The recent increase in the number of natural disasters, as well as terrorist attacks, has drawn much attention to the vulnerability of supply chains. Having passed through many links of the logistics chain that cover continents and organizations, the reliability and timeliness of cargo delivery is becoming increasingly difficult to assess. In particular, in seaports that are an indispensable hub in global supply chains and where complex sea and land interfaces take effect, the role of ports in disrupting marine supply chains needs to be explored. Loh, H. and Vinh, T. are examining the changing functions of the ports and supply chain helps determine the consequences and develop strategies to mitigate the effects of such disruptions [11].

The cooperation of carriers also helps in case of possible problems. The most frequently discussed operational risks of the port are port accidents [12,13], port equipment failures [13], improper handling of dangerous goods [14], port congestion [15], low qualification of personnel [16], safety violations [13], and strikes. In addition, human factors that impede communication (cultural differences, political problems, and conflicts between staff) may increase the scope of problems [17].

One of the main ways to quickly respond to the resulting failure is to change the logistics chain. Well-established schemes of vertical and horizontal cooperation of chain participants can help minimize the consequences: redirect goods to another node of the chain (another port) or change the type of transport.

Of course, one of the main issues is the sustainability of such coalitions. It, in turn, depends on the profitability of the participation of ports in the union. Here, the important role is played by the principle on which the profit of the coalition members is distributed.

At the modern market of sea shipping, there are several port alliances [3]. Examples include the following alliances:

1. Ports of Elba. In 2009, the ports of Cuxhaven, Brunsbüttel, Glückstadt, Stade, and Hamburg merged. The seaports of the three different federal states of Germany collaborate in the areas of marketing, customer analysis, and infrastructure planning and management;
2. Ports of the Rhine. The cross-border cooperation "RheinPorts" between the inner ports of the Rhine Basel in Switzerland, Mulhouse in France, and Weil am Rhein in Germany, began in 2007. The main joint tasks of neighboring ports in the three countries include service marketing and exchange of information, as well as handling of goods, customs, and container repair services. Since 2016, a joint logistics flow management system has been in use. The information system provides detailed information on the arrival of ships and connects various port participants, such as terminal operators, gateways, and transport companies;
3. Malmö–Copenhagen. In 2001, the ports of Malmö, Sweden and Copenhagen, Denmark agreed on the maximum possible form of cooperation—a joint venture responsible for cargo handling and storage. The ports are located in geographical proximity and benefit from direct and joint navigational access to the Oresund region. A joint venture called "Copenhagen Malmö Port CMP" is registered in Sweden. The company acts as a port and terminal operator in both cities. The objectives of cooperation are to focus on the various transport segments and to manage traffic at close range. The port of Copenhagen concentrates on imports and cruise shipments. The port of Malmö mostly serves as a transit cargo center.

2.3. Cooperation of Shipping Lines and Ports

The shipping line and the port are two different elements of the logistics chain and there is no direct competition between them. These conditions make the coalition easier and more profitable. Vertical integration of the shipping line and the port can effectively satisfy maritime logistics and increase the level of service. Shipping lines choose landing terminals for their ships; therefore, it becomes increasingly important for the port to cooperate with one or several shipping lines in order to ensure long-term prosperity. Shipping lines take a series of measures to increase efficiency and reduce costs. Global strategic alliances of sea carriers using large ships have appeared [18]. Cooperation with a large shipping line can guarantee port loading for a considerably long period of time.

The advantage of creating these kind of coalitions can be obtained in various ways. First, the shipping line may receive either a reduction in tariffs for cargo service, or a dedicated site where the shipping company is to be serviced without delay. In turn, shipping companies invest in the development of the port. In some cases, shipping companies have a share in the port.

2.4. Coalition of Shipping Lines

There is a large amount of research examining coalition of maritime carriers. For carriers, it is more advantageous to use large vessels. Therefore, they may combine to ensure full loading. At present, the largest alliances of maritime carriers have been formed in the world trade.

Various forms of cooperation help shipping lines expand their business, streamline their services, and reduce their costs in varying degrees [19]. The most demonstrative examples in history are price agreements between shipping lines [20]. There is also an agreement on the fleet and route sharing [21].

Currently, there are several major alliances of shipping companies: A.P. Molle–Maersk Group, Mediterranean Shipping Company S. A., CMA CGM Group, China Ocean Shipping Company (COSCO), Evergreen Marine, etc. Today, a more comprehensive and flexible form of cooperation has emerged, namely the global strategic alliance [19]. Panayides and Wiedmer in [22] explore the motivation to cooperate, naming among them the opportunity to share risks, eliminate duplicate routes, and use super-large vessels.

Shipping lines cooperation can be divided into a strategic or global alliance, an agreement on the ships sharing, and slot charters [22]. By sharing risks as well as improving customer service, product quality and market accessibility, partners in a global alliance gain a number of competitive advantages over others, thus increasing profitability [19].

2.5. Cooperation and Coopetition

Researchers also point out a new form of interaction between companies—coopetition. This is both cooperation and competition at the same time. Agreeing to cooperate on one issue, companies continue to compete in other areas. Dung-Ying Lin [23] offers a theoretical framework for characterizing cooperation and competition in international maritime shipping, and explores how carriers can manage their business models. In [24], various forms of cooperation and competition in the field of linear transportation, industrial and economic reasons for cooperation, and consequences for port competition in Europe are considered.

3. Model Conceptualization

The goal of this study is to analyze the profitability of the ports and land carriers' cooperation. Let us consider a large port hub (Seagate) and i peripheral ports, $i = 1, \ldots, n$. They serve the area with j customers, $j = 1, \ldots, m$. In this case, upon receiving an order, its delivery can be carried out in two ways. Cargo can be delivered by land immediately to the port hub (Figure 2).

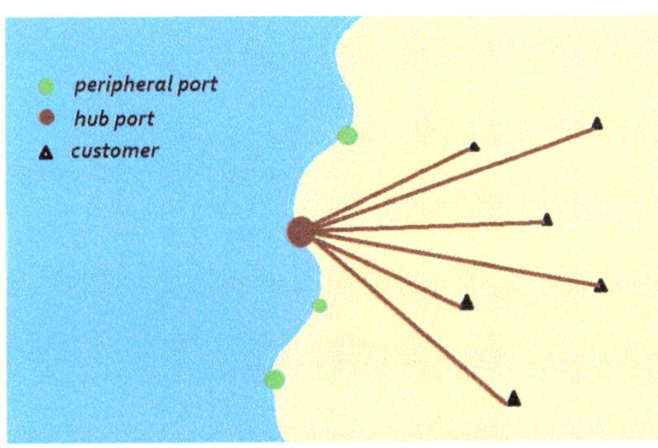

Figure 2. Ports network without cooperation.

It can also be delivered by land to the nearest peripheral port, and then by water to the hub. Further, the following notation is introduced:

q_j is the number of containers in the order from j. To simplify the model, tt is assumed that the cargo consists of standard containers;

p_i is the price assigned by port I for handling one container;

l_j^i is the overland distance between customer j and peripheral port i;

l_i^h is the distance from a peripheral port i to the hub port;

p_j^h is the overland distance from customer j to the hub;

$c(q_j, l_j^i)$ is the cost of shipping q_j containers over a distance of l_j^i by land;

$s(q_j, l_i^h)$ is the cost of shipping q_j containers over a distance of l_i^h by water;

V_i is the capacity of the port i (in containers);

V_i^*—currently available capacity of the i-th port;

m_j—customer losses in case of container demurrage (per day) or additional time of cargo handling in the peripheral port. In some cases, this value can be neglected. For example, when shipping goods over long distances, one or two additional days do not play a major role;

t_i—for a fixed time moment under consideration, the amount of time remaining until the peripheral port is ready to accept cargo for handling.

It is assumed that the hub port has a large capacity, and it is always available (waiting time is less than one day). Shipping costs when using the peripheral port are as follows:

$$TS_i^p = c(q_j, l_j^i) + s(q_j, l_i^h) + q_j p_i. \tag{1}$$

Costs in case of shipping directly to the hub:

$$TS^h = c(q_j, p_j^h). \tag{2}$$

Experience shows that in the process of delivery and handling of goods, disruptions are possible, leading to delays in the delivery of goods. These can be short delays associated with heavy traffic. Long delays are also possible, due to equipment breakdown, unfavorable weather conditions, etc. The cost of the delay includes the cost of lost profits for the transport company, the cost of waiting (idle time) at the peripheral port and the hub port, the cost of re-forming the cargo handling queue in ports, the cost of the route delay. The cost of delay per unit of time depends on many factors appropriate to a particular situation. Its assessment requires the work of experts. The duration of the delay is also an important factor. The probable duration of disruption is estimated on the basis of historical data by experts. Estimating the cost of delay is beyond the scope of this study. It is also necessary to take into account the additional time for handling cargo at the peripheral port. If necessary, one can also consider the possibility of marine shipping taking longer than overland delivery. This can be important, for example, in case of delivering perishable goods. This time difference can be accounted for in the t_i. variable.

Let us introduce the following notation:

t_{dis}—possible duration of the disruption;

t_{ex}—possible duration of the cargo processing in the peripheral port i.

Then, Formula (1) for the shipping cost is estimated as follows:

$$\begin{gathered} TS_i^p = c(q_j, l_j^i) + s(q_j, l_i^h) + q_j p_i + \alpha_i t_{dis} m_j + \beta_i t_{ex} \\ TS^h = c(q_j, p_j^h) + \alpha_h t_{dis} m_j. \end{gathered} \tag{3}$$

Variables α_h, α_i, and β_i indicate whether there is a problem on path i or h:

$$\alpha_i = \begin{cases} 1, & if\ there\ is\ a\ problem \\ 0, & if\ there\ is\ no\ problem \end{cases} \tag{4}$$

$$\begin{gathered} \alpha_h = \begin{cases} 1, & if\ there\ is\ a\ problem\ on\ the\ way\ to\ the\ hub \\ 0, & if\ there\ is\ no\ problem \end{cases} \\ \beta_i = \begin{cases} 1, & if\ there\ is\ a\ problem\ on\ the\ way \\ 0, & if\ there\ is\ no\ problem \end{cases} \end{gathered} \tag{5}$$

Let us propose the following algorithm for finding the optimal delivery method (optimal delivery route). The proposed algorithm takes into account, not only the distances between customers and ports, but also the feasibility of using an intermediate peripheral port, as in this case the goods go through additional processing, which requires additional expenditures. The algorithm also takes into account the current situation on the roads and in ports. If obstacles arise on a certain route, the feasibility of switching to another route is assessed. To do this, the losses from waiting for the restoration of the original supply chain and the possible additional costs when changing the route (i.e., the cost of handling cargo at an intermediate port or an increased distance) are compared.

The purpose of the algorithm is to find the optimal (minimum cost) cargo delivery route. For this, the cost of delivery only by land transport directly to the hub is compared with the cost of a route that includes sea transportation with the participation of one of the intermediate ports. Of course, it is necessary to take into account the additional costs of handling cargo at the intermediate port. However, shipping by sea is several times cheaper, which will save on costs.

The following Algorithm 1 of decision making is proposed:

Algorithm 1 An optimal route search algorithm.

1. An order arrives from the customer j;
2. All sections of the route and ports are checked. If there is a disruption on some section of the route or in the peripheral port, then for this section the cost of delivery is estimated using Formulas (1).
3. For all i, the following values are calculated:
4. TS_i^p, TS^h
5. $TS^{\min} = \min_i \left(TS_i^p, TS^h \right)$
6. If $TS^{\min} = TS^h$, then the cargo is shipped by land directly to the port hub.
7. If $TS^{\min} = TS_i^p$ for some i, then it is checked that this port is currently available: $q_j < V_i^*$.
8. If the condition $q_j < V_i^*$ is not satisfied, then for the obtained i the following value is defined:
9. $TS_i^p = c(q_j, l_j^i) + s(q_j, l_i^h) + q_j p_i + m_j t_i$.

The algorithm runs again from the second step. The algorithm stop criterion is selected by the algorithm running time or by the number of iterations.

Cooperative Game

With each order, a path is determined that minimizes the total costs of the coalition of carriers and customers and, therefore, increases profits. The next problem is how to divide the total profit between the members of the alliance. The costs of each individual carrier are allocated according to the Shapley vector. The Shapley vector was chosen based on its properties. As E. Moulin notes, Shapley's vector is based on sequential accounting of additional income from joining a fixed participant to each coalition [25]. This property is important, as the members of such a coalition are "unequal", and a large port hub contributes a significant part of the coalition's profit. For example, let us consider the alliance of three ports. Consider a cooperative game $\Gamma(N, v)$ with three players. Here, $N = \{1,2,3\}$, where {1} is a hub port, and {2} and {3} are peripheral ports. Grand coalition is {1,2,3}. The characteristic function describing the profit of the grand coalition is denoted by $v(\{1,2,3\})$. Characteristic functions $v(\{1,2\})$, $v(\{1,3\})$ are also specified. As the main goal is to deliver the cargo to the hub, coalitions without a hub have no profit:

$$v(\{2\}) = v(\{3\}) = v(\{2,3\}) = 0 \tag{6}$$

The alliance members divide the received profit in accordance with the Shapley vector, where φ_i is the profit (gain) of the i-th player.

For the game with three ports the following profits are obtained:

$$\begin{aligned}
\varphi_2 &= \tfrac{1}{3}(v(1,2,3) - v(2,3)) + \tfrac{1}{6}(v(1,2) - v(2)) + \tfrac{1}{6}(v(1,3) - v(3)) \\
\varphi_1 &= \tfrac{1}{3}(v(1,2,3) - v(1,3)) + \tfrac{1}{6}(v(1,2) - v(1)) + \tfrac{1}{6}(v(2,3) - v(3)) \\
\varphi_3 &= \tfrac{1}{3}(v(1,2,3) - v(1,2)) + \tfrac{1}{6}(v(1,3) - v(1)) + \tfrac{1}{6}(v(2,3) - v(2))
\end{aligned} \tag{7}$$

4. Results and Discussion

Numerical experiments were carried out to demonstrate the performance of the algorithm. As an example, the region with one main and three peripheral ports was considered (see Figure 3). Ports receive delivery orders. Port #1 is the main port, and all shipments go through it. Possible delivery options are either by land transport directly to port #1, or through a peripheral port. In the latter case, a part of the way is provided by land transport to the peripheral port, and then by water transport to port #1. The parameters of the demand function were varied, and the equilibrium competition and cooperation levels were estimated. For numerical simulation, the parameters were varied in the following intervals: $q_j = [20, 150]$, $p_i^h = [20; 100]$, $c = [1.5; 3]$, $s = [0.3; 1]$, $l_i^h = [50; 500]$, $p_i = [150; 350]$, $t_i = [1; 3]$, and $m_j = 500$. The algorithm was implemented in MATLAB.

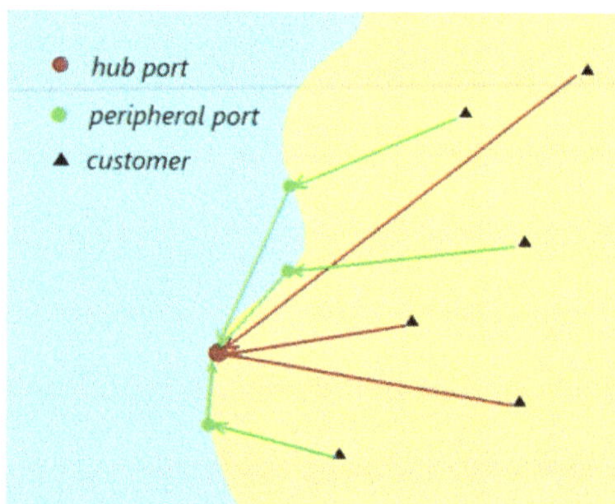

Figure 3. Ports network example.

Sample results of the numerical simulations are presented in Table 1.

Table 1. Sample results of the numerical simulations.

Port # That Received the Order	Port # That Handled the Order	Costs without Cooperation	Costs with Cooperation	Disruption	Savings%
1	1	11,624	11,624	no	0.00%
1	1	23,008	23,008	no	0.00%
1	4	12,000	11,548	yes	3.77%
3	4	17,968	16,843	yes	6.26%
1	2	9600	8800	no	8.33%
4	3	28,000	25,000	no	10.71%
2	2	13,654	13,654	no	0.00%
3	4	28,345	27,567	no	2.74%
4	3	23,780	22,445	yes	5.61%
3	1	21,658	19,567	no	9.65%
3	3	7846	7846	no	0.00%
2	1	6810	6654	yes	2.29%
3	1	22,365	20,248	no	9.47%
2	2	13,568	13,568	no	0.00%

The data in Table 1 contain information about the port that received the order, the port to which the cargo was redirected after determining the optimal route, the initial cost of delivery of the cargo, and the cost after determining the optimal route, taking into account possible disruptions. The main idea in selected examples was to observe the percentage of successful and unsuccessful coalitions. Numerical experiments showed that linking ports in order to reduce transport costs and losses in the event of network outages helped to reduce costs by an average of 5%. This study does not take into account the situation where different ports and land carriers belong to different alliances. Our goal is to demonstrate the benefits of cooperation and highlight the value of peripheral ports, many of which are on the verge of closure.

5. Conclusions

In response to strong competition in the container shipping markets, collaboration became a major focus of this era. Consolidation of sea routes, globalization of shipping lines, and cooperation of port operators emerged. The presented study uses a game-theoretic approach to model port consolidation, and describes a multi-port game scenario. The presented model assumes horizontal cooperation of ports. Restrictions on alliances were not introduced deliberately, and cooperation between alliances was considered. The cooperation of terminal operators, located in close proximity to one another, improves the quality of services for the main customers—liner carriers—and helps to reduce transport costs. Among the expected positive effects of the considered cooperation are more efficient use of existing port infrastructures through equipment sharing, and flexible workforce distribution, as well as better rationalization of traffic peaks.

The study demonstrated the effectiveness of cooperation between ports and land carriers in reducing the effects of disruptions as well as reducing overall logistics costs. Numerical experiments have shown that the efficiency of such cooperation is 5% on average. In some cases, the savings are more than 10%.

Cooperation of sea carriers, land carriers, and ports, uniting peripheral ports and hubs in a single coalition, gives such alliances the opportunity to increase the level of loading and unloading services, and allows sea routes to receive a stable income from the port operation. It also protects peripheral ports from closure, saving jobs and financial revenues to economic regions. Of course, such a tactic is not a panacea, and there are cases when the savings from such cooperation cover the losses for unclaimed peripheral ports.

An important issue is the stability of such coalitions. The creation of such alliances involves formal negotiations, signing agreements, and can be quite costly. The issue of coalition stability may be a topic for future research.

Author Contributions: Conceptualization, E.A.L. and Y.E.B.; methodology, E.A.L. and Y.E.B.; writing—original draft preparation, E.A.L. and Y.E.B.; and writing—review and editing, E.A.L. and Y.E.B. All authors have read and agreed to the published version of the manuscript.

Funding: This research received no external funding.

Institutional Review Board Statement: Not applicable.

Informed Consent Statement: Not applicable.

Conflicts of Interest: The authors declare no conflict of interest.

References

1. United Nations Conference on Trade and development. Review of Maritime Transport 2010. Report by the UNCTAD Secretariat. United Nations. 2010. Available online: https://unctad.org/system/files/official-document/rmt2010_en.pdf (accessed on 7 January 2021).
2. Počuča, M.; Zanne, M. The impact of global economic crisis on the dry bulk shipping industry. *Pomorstvo* **2009**, *23*, 477–493.
3. Fraunhofer Center for Maritime Logistics and Services: Port Cooperation between European Seaports—Fundamentals, Challenges and Good Practices. The Report. Available online: www.guengl.eu/issues/publications/port-cooperation-between-european-seaports (accessed on 15 December 2020).
4. De Martino, M.; Marasco, A.; Morvillo, A. Supply Chain Integration and Port Competitiveness: A Network Approach. In *Supply Chain Innovation for Competing in Highly Dynamic Markets: Challenges and Solutions*; Pietro, E., Ed.; IGI Global: Hershey, PA, USA, 2012; pp. 62–77. [CrossRef]
5. Zhaojing, W.; Hao, H.; Qingcheng, Z.; Xiaobing, L. Profit Sharing and the Stability of Shipping Alliances Based on Game Theory. *J. Transp. Econ. Policy* **2016**, *50*, 245–261. [CrossRef]

6. Fernando, Y.; Jasmi, M.F.A.; Shaharudin, M.S. Maritime green supply chain management: Its light and shadow on the bottom line dimensions of sustainable business performance. *Int. J. Shipp. Transp. Logist.* **2018**, *11*, 60–93. [CrossRef]
7. Chul-hwan, H. Assessing the impacts of port supply chain integration on port performance. *Asian J. Shipp. Logist.* **2018**, *34*, 129–135.
8. Notteboom, T.; Rodrigue, J. Re-Assessing Port-Hinterland Relationships in the Context of Global Commodity Chains. In *Ports, Cities, and Global Supply Chains*; Routledge: Oxfordshire, UK, 2007; pp. 67–82. [CrossRef]
9. Lezhnina, E.A.; Zakharov, V.V. The Nash equilibrium in multy-product inventory model. *Contrib. Game Theory Manag.* **2014**, *7*, 191–200.
10. Ang, S.H. Competitive intensity and collaboration: Impact on firm growth across technological environments. *Strateg. Manag. J.* **2008**, *29*, 1057–1075. [CrossRef]
11. Loh, H.; Vinh, T. The Role of Ports in Supply Chain Disruption Management. In Proceedings of the International Forum on Shipping, Ports and Airports (IFSPA) 2012: Transport Logistics for Sustainable Growth at a New Level, Hong Kong, China, 27–30 May 2012.
12. Pinto, C.A.; Talley, W.K. The Security Incident Cycle of Ports. *Marit. Econ Logist.* **2006**, *8*, 267–286. [CrossRef]
13. Mennis, E.; Platis, A.; Lagoudis, I.; Nikitakos, N. Improving Port Container Terminal Efficiency with the use of Markov Theory. *Marit. Econ Logist.* **2008**, *10*, 243–257. [CrossRef]
14. Ellis, J. Analysis of accidents and incidents occurring during transport of packaged dangerous goods by sea. *Saf. Sci.* **2011**, *49*, 1231–1237. [CrossRef]
15. Paul, J.A.; Maloni, M.J. Modeling the effects of port disasters. *Marit. Econ Logist.* **2010**, *12*, 127–146. [CrossRef]
16. Fabiano, B.; Curro, F.; Reverberi, A.P.; Pastorino, R. Port safety and the container revolution: A statistical study on human factor and occupational accidents over the long period. *Saf. Sci.* **2010**, *48*, 980–990. [CrossRef]
17. Horck, J. Cultural and Gender Diversities Affecting the Ship/Port Interface—Maritime Education and Training Efforts to Bridge Diversity Gaps. In Proceedings of the 1st International Ship-Port-Interface Conference—The Human Element (ISPIC), Bremen, Germany, 19–21 May 2008.
18. Asteris, M.; Collins, A. UK container port investment and competition: Impediments to the market. *Transp. Rev.* **2010**, *30*, 163–178. [CrossRef]
19. Bergantino, A.S.; Veenstra, A.W. Interconnection and Co-ordination: An Application of Network Theory to Liner Shipping. *Int. J. Marit. Econ.* **2002**, *4*, 231–248. [CrossRef]
20. Tang, O.; Sun, P. Anti-competition of ocean shipping alliances: A legal perspective. *Marit. Bus. Rev.* **2018**, *1*, 4–19. [CrossRef]
21. The Impact of Alliances in Container Shipping. Available online: www.itf-oecd.org/impact-alliances-container-shipping (accessed on 15 November 2020).
22. Panayides, P.; Wiedmer, R. Strategic alliances in container liner shipping. *Res. Transp. Econ.* **2011**, *32*, 25–38. [CrossRef]
23. Lin, D.-Y.; Huang, C.-C. Coopetition in international maritime shipping. *Pol. Marit. Res. Spec. Issue* **2013**, *20*, 5–12. [CrossRef]
24. Heaver, T.; Meersman, H.; Moglia, F.; Van de Voorde, E. Do Mergers and Alliances Influence European Shipping and Port Competition? *J. Marit. Policy Manag.* **2000**, *27*, 363–373. [CrossRef]
25. Moulin, H. *Axioms of Cooperative Decision Making (Econometric Society Monographs)*; Cambridge University Press: Cambridge, UK, 1988. [CrossRef]

Journal of
Marine Science and Engineering

MDPI

Article

Multicriteria Ship Route Planning Method Based on Improved Particle Swarm Optimization–Genetic Algorithm

Wei Zhao, Yan Wang, Zhanshuo Zhang and Hongbo Wang *

State Key Laboratory on Integrated Optoelectronics, College of Electronic Science and Engineering, Jilin University, Changchun 130000, China; weizhao18@mails.jlu.edu.cn (W.Z.); yanwang20@mails.jlu.edu.cn (Y.W.); zszhang19@mails.jlu.edu.cn (Z.Z.)
* Correspondence: wang_hongbo@jlu.edu.cn; Tel.: +86-0431-8514-8242

Abstract: With the continuous prosperity and development of the shipping industry, it is necessary and meaningful to plan a safe, green, and efficient route for ships sailing far away. In this study, a hybrid multicriteria ship route planning method based on improved particle swarm optimization–genetic algorithm is presented, which aims to optimize the meteorological risk, fuel consumption, and navigation time associated with a ship. The proposed algorithm not only has the fast convergence of the particle swarm algorithm but also improves the diversity of solutions by applying the crossover operation, selection operation, and multigroup elite selection operation of the genetic algorithm and improving the Pareto optimal frontier distribution. Based on the Pareto optimal solution set obtained by the algorithm, the minimum-navigation-time route, the minimum-fuel-consumption route, the minimum-navigation-risk route, and the recommended route can be obtained. Herein, a simulation experiment is conducted with respect to a container ship, and the optimization route is compared and analyzed. Experimental results show that the proposed algorithm can plan a series of feasible ship routes to ensure safety, greenness, and economy and that it provides route selection references for captains and shipping companies.

Keywords: multicriteria route planning; genetic algorithm; particle swarm optimization; oceanic meteorological routing

Citation: Zhao, W.; Wang, Y.; Zhang, Z.; Wang, H. Multicriteria Ship Route Planning Method Based on Improved Particle Swarm Optimization–Genetic Algorithm. *J. Mar. Sci. Eng.* **2021**, *9*, 357. https://doi.org/10.3390/jmse9040357

Academic Editor: Alejandro J. C. Crespo

Received: 28 February 2021
Accepted: 19 March 2021
Published: 25 March 2021

1. Introduction

With the progress of navigation technology, the safety and energy-saving problems associated with maritime navigation have gradually become the focus of human attention. Severe winds, waves, and other meteorological factors seriously affect the safety of ships when sailing at sea. Currently, the weather forecasting technology has rapidly developed. The weather forecast information can be used to plan routes for ships to avoid severe winds and waves. Recently, the International Maritime Organization (IMO) and governments have paid close attention to the air pollution and energy consumption of ships. In 2018, the IMO adopted the "preliminary strategy for greenhouse gas emission reduction of IMO ships," thereby sending a strong signal to the international community that the shipping industry is turning into a low-carbon industry [1]. The reduction of fuel consumption and carbon emissions through reasonable route planning is an important measure in response to the low-carbon strategy. To plan an efficient route, the captain must completely utilize the weather forecast data according to the voyage mission and characteristics of the ship and avoid selecting high-risk routes against winds and waves. The following elements must be fully considered when planning multicriteria ship routes.

- The navigation time of the ship must be reduced.
- The fuel consumption of ships must be reduced.
- Areas with severe sea conditions must be avoided, ship rolling must be reduced, and the safety of ships and cargo must be ensured.

In the case of the first element, regardless of ship stalls and weather risks, the great circle route is an ideal route because of the short voyage, thereby reducing the navigation time [2]. The main method for realizing the second element is to select the economic route. The main method for realizing the third element is to select safe routes that avoid areas with high meteorological risks based on weather forecast data and ship characteristics. The prosperity and development of shipping can be further promoted by finding an appropriate methodology to comprehensively consider the three elements to design reasonable, safe, and green alternative routes for ships.

Currently, with the rapid development of the optimization theory, some single-criteria and multicriteria route planning algorithms for ships have been proposed. Initially, some traditional mathematical methods were applied to solve the problem of ship route planning.

For example, James applied the isochron method to solve the problem of ship route planning under meteorological conditions [3]. However, this method has the "isochron loop" problem, making it unsuitable for computer-aided calculation. To solve this problem, Hagiwara et al. proposed a modified isochron method [4], and Lin et al. proposed a three-dimensional modified isochron method [5]. These two methods consider minimum fuel consumption and minimum navigation time as optimization goals. The route is optimized, but there is still a problem of complex calculations. Smierzchalski et al. used the isochron method for generating the initial route and used the evolutionary algorithm to obtain the optimal ship route [6]. Shao et al. proposed a forward, three-dimensional, dynamic planning algorithm and planned the route intending to minimize fuel consumption [7]. Sen [8] used Dijkstra's algorithm to solve the multicriteria route planning problem of ships, focusing on the optimization goals of navigation time. Mannarini et al. [9] used a graph-search method with time-dependent edge weights to optimize ship routes. The optimal route may be longer in terms of miles sailed, and yet it is faster and safer than the geodetic route between the same departure and arrival locations. With the continuous development of intelligent optimization algorithms and big data, these technologies are gradually being used to solve the problem of ship route planning. For example, Wang et al. used real-coded genetic algorithms to plan ship routes with the goal of minimizing navigation time and risk [10]. Chuang et al. applied the fuzzy genetic method by considering the transportation and berthing time of container ships and planned ship routes [11]. Wang et al. considered ship maneuverability and applied a double-loop genetic algorithm to achieve dynamic path planning for ships [12]. Vlachos used a simulated annealing algorithm to plan the optimal ship path based on predicted wind and wave data [13]. Tsou used the ant colony algorithm and genetic algorithm to plan the route of ships with minimum fuel consumption [14]. Zhang et al. proposed an improved multiobjective ant colony algorithm by considering navigation time and navigation risk as the optimization goals and performed ship route planning [15]. Vettor et al. applied a multiobjective evolutionary algorithm to plan the best weather route for ships [16,17]. He et al. generated an optimized route for the ship based on historical automatic identification system (AIS) data. Although the length of the generated route is slightly smaller than the actual ship's trajectory, in a complicated environment, some routes may cross obstacles or be in shallow waters [18]. The combination of particle swarm optimization and genetic algorithm has a good effect on solving the route optimization problems. Abd-El-Wahed et al. verified the superiority of the combination of particle swarm optimization (PSO) and genetic algorithm (GA) in solving nonlinear optimization problems [19]. Liu et al. proposed a method in which GA and PSO were combined to solve route planning problems in restricted waters with a single optimization goal [20].

Although various solution algorithms have been developed for route planning problems, majority of them are based on a single criterion or on two criteria without considering more optimization goals and providing more routes for selection. To solve this problem, this study proposes a multicriteria ship route planning method based on an improved PSO–GA to provide more route options for captains and shipping companies. The research content and structure of this paper are as follows.

First, this study establishes the framework of multicriteria route planning based on the elements of multicriteria route planning and ship navigation characteristics. Second, a mathematical model related to shipping route design is established, including mathematical models of navigation time, fuel consumption, and navigation risk. Then, considering the static constraints (coastlines, islands, and reefs) and dynamic constraints (severe wind and wave areas) observed at sea, the multicriteria PSO–GA is proposed to solve the route planning problem. Here, particle cooperative operations are used to improve the convergence speed of the algorithm. Further, crossover, mutation operations, and multigroup elite selection operations are used to enhance the diversity of the population. The improved Pareto frontier solution selection strategy is used to avoid the algorithm from falling into local optimality. Finally, the recommended route selection criteria are proposed based on the Pareto optimal frontier and Pareto optimal solution set obtained using the algorithm. A simulation experiment is designed for a container ship. Experimental results show that the proposed algorithm can plan the route to ensure safety, economy, and time savings and that it can provide a recommended route as well as a series of alternative routes.

2. Methods

2.1. Multicriteria Route Planning Framework

This study establishes a multicriteria route planning framework to introduce the structure of this study more clearly. As shown in Figure 1, the framework consists of six parts, i.e., optimization criteria, ship speed analysis, model construction, multicriteria algorithm, route evaluation, and route selection.

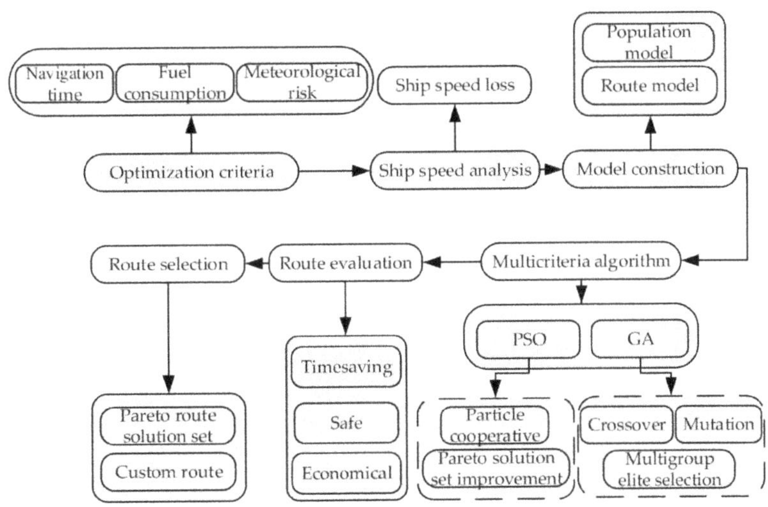

Figure 1. Multicriteria route planning framework.

The optimization criteria include the navigation time, meteorological risk, and fuel consumption. The navigation time can be obtained by adding the time associated with each route segment. Further, meteorological risks are numerically processed according to the potential risks caused by wind and waves to the ship. Fuel consumption is simulated by parameters such as rated power and ship speed. The speed loss of the ship under wind and wave conditions is analyzed in the ship speed analysis part. The model construction part includes construction of the route model and the population coding method. The multicriteria algorithm includes particle cooperative operation, crossover operation, mutation operation, multigroup elite selection, and improved Pareto solution set distribution method. The main evaluation criteria of the route evaluation part include time savings, safety, and economy. Route selection includes two parts. One part is to provide multiple route solution sets obtained via algorithm optimization, whereas the other part aims to provide the recommended route that best meets the requirements based on custom target values.

2.2. Optimization Criteria

It is necessary to establish guidelines for ship route optimization to reflect the quality of ship routes. Further, different objective functions must be designed according to the optimization criteria to evaluate the performance of the route. This study uses the following three variables to measure the performance of routes: navigation time, meteorological risk, and fuel consumption.

2.2.1. Navigation Time

As shown in Figure 2, the route planned in this study includes a series of waypoints. Therefore, the total navigation time of the ship from the departure point to the target point can be obtained by adding the time spent on each route, as shown in Equation (1).

$$T_{total} = \sum_{i=0}^{n-1} t_i;\ t_i = \frac{L_i}{V_a^i} \tag{1}$$

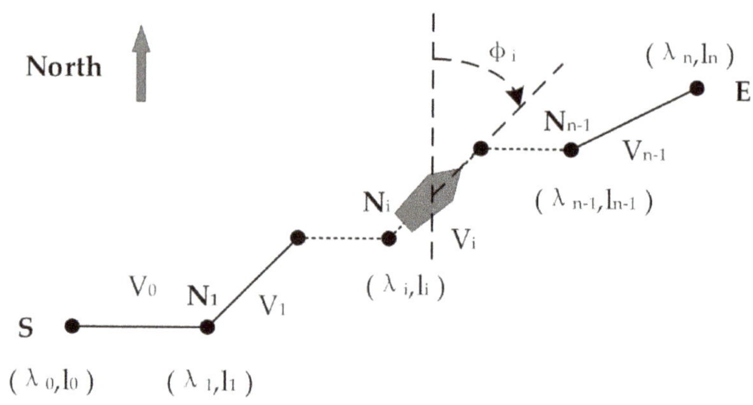

Figure 2. Route diagram comprising multiple waypoints.

Here, T_{total} is the total navigation time of the ship, t_i is the navigation time of the ship on each route section, and V_a^i is the actual speed of the ship on the ith route segment. If Earth is considered as an ellipsoid, the length of any two points on the Mercator projection map can be calculated as follows [10]:

$$\frac{l_2 - l_1}{\tan\varphi_{rh}} = \ln\left[\tan\left(\frac{\pi}{4} + \frac{\lambda_2}{2}\right)\left(\frac{1 - e\sin\lambda_2}{1 + e\sin\lambda_2}\right)^{\frac{e}{2}}\right] - \ln\left[\tan\left(\frac{\pi}{4} + \frac{\lambda_1}{2}\right)\left(\frac{1 - e\sin\lambda_1}{1 + e\sin\lambda_1}\right)^{\frac{e}{2}}\right], \tag{2}$$

$$L_{rh} = (\lambda_2 - \lambda_1) \cdot \sec\varphi_{rh}, \tag{3}$$

where λ_1 and l_1 are the latitude and longitude coordinates of the first point, respectively, λ_2 and l_2 are the latitude and longitude coordinates of the second point, respectively, φ_{rh} is the direction of the rhumb line, L_{rh} is the distance between two points (in radians), and e is the eccentricity of Earth. The above formula can be applied to ships sailing along the non-isolatitude lines. However, in the case of isolatitude lines, i.e., when the ship's heading is 90° or 270°, the following equation can be applied.

$$L_{rh} = (l_2 - l_1) \cdot \cos \lambda_1. \tag{4}$$

2.2.2. Meteorological Risk

It is necessary to understand the degree of threats to the safety of ship navigation caused by weather and ocean conditions. The IMO has given out information on how the captain should choose the route to avoid the navigation risk zone under severe sea conditions [21], but it did not assess the overall risk status of the route. In this section, we comprehensively consider wind, waves, and seakeeping to assess ship navigation risks and propose a comprehensive risk calculation formula to adapt to route optimization under good and severe sea conditions.

According to the "2008 International Intact Stability Regulations" [22], the stability criterion K should satisfy Equation (5) to ensure that ships navigate safely in strong winds.

$$K = \frac{L_q}{L_f} \geq 1, \tag{5}$$

where L_q represents the minimum overturning moment arm, which can be obtained from the dynamic stability curve and roll angle. L_f represents the wind pressure roll arm, and its value can be obtained using Equations (6) and (7).

$$L_f = \frac{P \cdot A_f \cdot Z}{1000 \cdot g \cdot \Delta}, \tag{6}$$

$$P = \frac{C_p \cdot \rho \cdot u^2}{2}, \tag{7}$$

where P represents the unit calculated wind pressure, A_f represents the wind area of the ship, Z represents the height from the center of the wind area to the water surface, g represents the acceleration of gravity, Δ represents the ship displacement,C_p represents the wind pressure coefficient, ρ represents the air density, and u represents the average wind speed. Based on Equations (5)–(7), the crosswind speed that the ship can withstand 10 m above the sea surface should satisfy Equation (8).

$$u_{10} \leq u_{10\max} = 40 \cdot \left(\frac{10}{Z}\right)^{\frac{1}{8}} \cdot \sqrt{\frac{10 \cdot \Delta \cdot L_q}{C_p \cdot A_f \cdot Z}}, \tag{8}$$

According to the maximum crosswind that can be withstood by the ship, a numerical expression of the risk of wind to the ship is established, as shown in Equation (9), where u_{cross} represents the lateral wind speed experienced by the ship. The risk is a gradual process. We think that when the value is greater than 0.6, it is unacceptable. When it is less than 0.6, a route with as low a risk value as possible should be chosen.

$$risk_{wind} = \begin{cases} \frac{u_{cross}}{u_{10\max}}, & if\ the\ value\ is\ less\ than\ 1 \\ 1 \quad , & else \end{cases}. \tag{9}$$

Under severe weather conditions, rolling is an important factor that causes a ship to capsize [23]. Therefore, in this study, the risk value caused by waves is described according to the rolling of the ship. The encounter period between the ship and wave is shown in Equation (10).

$$T_E = \frac{\lambda}{\left|1.25 \cdot \sqrt{\lambda} + V \cdot \cos \mu\right|}, \tag{10}$$

where λ represents the wavelength, V represents the shipping speed, and μ represents the angle between the ship's motion direction and wave direction.

The natural rolling period, T_θ, of the ship can be calculated as follows:

$$T_\theta = \frac{2 \cdot C \cdot B}{\sqrt{GM}}, \tag{11}$$

where C represents the rolling period of the ship, B represents the width of the ship, and GM represents the height of initial stability.

According to the resonance theory of a ship in waves, the ship is in the harmonic rolling area when $0.70 < T_\theta / T_E < 1.3$. In this area, a ship may have a large roll angle, threatening its safety [23]. Therefore, we have established a numerical expression for the risk of waves to ships as follows.

$$risk_{wave} = \begin{cases} \frac{T_\theta}{T_E} &, \ if\, 0 \le \frac{T_\theta}{T_E} < 1 \\ 2 - \frac{T_\theta}{T_E}, & if\, 1 \le \frac{T_\theta}{T_E} < 2 \\ 0 &, \ \frac{T_\theta}{T_E} \ge 2 \end{cases}. \tag{12}$$

According to Equation (12), there is an absolute risk when $risk_{wave} > 0.7$, and when it is less than 0.7, the risk gradually decreases.

To study the ship's motion state in wind and waves, seakeeping is a factor that must be considered. There are many factors that affect seakeeping, such as the ship's roll, pitch, heave, the probability of the green water on deck, the probability of slamming occurrence, and bow vertical acceleration, etc. In this study, seakeeping is determined by the following three factors: the amplitude of the pitch motion, the probability of slamming occurrence, and the probability of green water on deck. The limit values of the pitch amplitude are based on the NATO STANAG, Standardization Agreement, 4154 criteria, slamming probability and green water on deck probability comply with the NORDFORSK 1987 criteria [24,25], as detailed in Table 1.

Table 1. General operability limiting criteria for ships.

Root mean square of pitch amplitude (RMS_{pl})	1.5 degrees
Slamming probability (P_{spl})	0.03 (L $\le$ 100 m) or 0.01 (L $\le$ 330 m)
Green water on deck probability (P_{wdl})	0.05

The ship's risk obtained by considering seakeeping is determined according to Equation (13) [26]:

$$risk_{seakeeping} = 1 - \max\left\{0; \left(1 - \frac{RMS_p}{RMS_{pl}}\right) \cdot \left(1 - \frac{P_{sp}}{P_{spl}}\right) \cdot \left(1 - \frac{P_{wd}}{P_{wdl}}\right)\right\}, \tag{13}$$

where RMS_p is the root mean square (RMS) of the pitch motion amplitude, P_{sp} is the probability of occurrence of slamming, P_{wd} is the probability of water on deck.

Based on the ship Response Amplitude Operator, the RMS of the pitch motion amplitude is determined according to the following equation [27,28]:

$$RMS_p = \sqrt{\int_0^\infty |H_5(\omega_e)|^2 \cdot S_\varsigma(\omega_e)}, \tag{14}$$

where H_5 is the speed-dependent pitch motion transfer function, S_ς is the wave spectrum, $\omega_e = \omega - \omega^2\psi$ is the encounter wave frequency that satisfies the Doppler shift equation, depending on the absolute wave frequency, ω, and the factor $\psi = U \cdot \cos(\mu/g)$, where the vessel speed is denoted by U and μ denotes the heading angle [29].

$$P_{sp} = e^{-(\frac{v_{cr}^2}{2C_s^2 \cdot m_{2,r}} + \frac{d^2}{2C_s^2 \cdot m_{0,r}})}, \tag{15}$$

where $v_{cr} = 0.093\sqrt{g \cdot L}$ is the threshold velocity, C_s is the swell up coefficient, and d is the ship draught at the forward perpendicular. $m_{0,r}$ and $m_{2,r}$ denote the zero-order and second-order spectral moments of the ship's relative motion as regards the sea surface [29].

$$P_{wd} = e^{-\frac{f_b^2}{2C_s^2 \cdot m_{0,r}}}, \tag{16}$$

where f_b is the freeboard at the ship forward perpendicular.

Based on the above analysis, we established the comprehensive risk of the ship being disturbed by winds and waves in the ith route segment as follows:

$$risk^i = a_1 \cdot risk^i_{wind} + a_2 \cdot risk^i_{wave} + a_3 \cdot risk^i_{seakeeping}, \quad \sum_{j=1}^{3} a_j = 1, \tag{17}$$

Because the navigation risk is a gradual process, when $a_i = 1/3$, we believe that when the risk value is greater than 0.6, the route will be discarded. When the route value is less than 0.6, although the route is acceptable, it is still better to keep the risk value is as small as possible. The risk distribution of the entire route is

$$risk = \left\{risk^0, risk^1, \cdots, risk^{n-2}, risk^{n-1}\right\}. \tag{18}$$

Therefore, the total risk of a route is

$$RISK = \max(risk). \tag{19}$$

2.2.3. Fuel Consumption

The total fuel consumption associated with each ship route can be obtained by adding the fuel consumption in multiple route sections. The total fuel consumption of a ship can be given as follows:

$$f_{fuel} = \sum_{i=1}^{m-1} (Q_{ti} \cdot t_i). \tag{20}$$

The fuel consumption of a ship during navigation is related to many factors such as the main engine structure, ship type structure, loading capacity, sailing speed, fuel type, and sea conditions. The best way to develop a reliable speed–consumption function is by collecting real data about the speed and the corresponding consumption [30]. The speed–consumption function is described as an exponential function, with vessel design as a constant and speed as a variable in the exponent. In this paper, Euler's number, e, was taken as the basis to simplify later calculation. The ship's fuel consumption per unit time can be represented as follows:

$$Q_{ti} = a \cdot e^{b \cdot v}, \tag{21}$$

where a and b are parameters calculated for each vessel by exponential regression. The total fuel consumption of the ship can be approximated by Equations (20) and (21).

2.3. Ship Speed Loss

When a ship is sailing, winds and waves will cause additional resistance. The actual speed of the ship under winds and waves will be lower than the speed in still water when keeping the ship's propulsion power constant [31]. The speed loss of the ship will affect the navigation time and fuel consumption and considerably affect the results of the ship's multicriteria route planning. Therefore, speed loss is a factor that must be considered during route planning.

The following methods are mainly used to estimate the ship's speed loss. The first method is theoretical derivation calculation, based on which the actual speed of the ship can be estimated from the perspective of system energy conservation. However, this method is cumbersome and difficult to calculate. The second method is the test method, which uses equipment, such as pools and wind tunnels, to simulate the ship for determining the speed loss. However, the overall applicability of this method is poor. The third method is an experimental method, wherein a large amount of actual observation data is used. After statistical analysis, an empirical formula is obtained for estimating the stall. Because the third method is conducive for the introduction of the algorithm presented in this study, the calculation formula of ship speed loss proposed by Feng was used here [32].

$$V_a = V_0 - \left(1.08 \cdot h - 0.126 \cdot q \cdot h + 2.77 \cdot 10^{-3} \cdot F \cdot \cos\alpha\right) \cdot \left(1 - 2.33 \cdot 10^{-7} \cdot D \cdot V_0\right). \quad (22)$$

where V_a is the actual speed of the ship under winds and waves, V_0 is the hydrostatic speed of the ship, F is the wind speed, D is the actual displacement of the ship, h is the significant wave height, q is the relative angle between the ship's heading and wave direction, and α is the relative angle between the ship's heading and wind direction.

According to the above analysis, when the ship sails in rough weather, the ship will have slamming occurrence and the green water on the deck. In order to ensure the safety of navigation, the captain will take the initiative to reduce the speed of the ship. Therefore, seakeeping must be considered to calculate the maximum speed allowed in wind and waves. In this study, when the maximum speed is exceeded, the probability of the route being selected decreases. The maximum allowable speed is determined as follows [10].

$$V_c = \exp\left[0.13 \cdot (\mu(q) - h)^{1.6}\right] + r(q), \quad (23)$$

where $\mu(q) = 12.0 + 1.4 \cdot 10^{-4} \cdot q^{2.3}$, $r(q) = 7.0 + 4.0 \cdot 10^{-4} \cdot q^{2.3}$.

2.4. Population Model

The population of multicriteria PSO–GA comprises multiple individuals, and each individual is represented by a series of latitude and longitude coordinates. In Equation (19), a shipping route is represented, where X_i is a two-dimensional vector containing longitude and latitude values.

$$\boldsymbol{X} = [X_0, X_1, \cdots X_i \cdots, X_{n-1}, X_n]. \quad (24)$$

Each route can be generated in a limited search area based on the reference route. The reference route is the high-frequency route of the ship in previous voyage missions. The limited search area is the area expanded on both sides of the reference route based on the historical route characteristics of the ship. As shown in Figure 3, S and E are the starting and ending points, respectively, the reference route is the great circle route between these two points, and the area enclosed by the dash-dotted line is the search area of the ship waypoint. The upper boundary of the search area is UB, whereas the lower boundary is LB, which are represented by Equations (25) and (26), respectively.

$$UB = \{Up_0 \cdots Up_i \cdots Up_n\}, \quad (25)$$

$$LB = \{Low_0 \cdots Low_i \cdots Low_n\}, \quad (26)$$

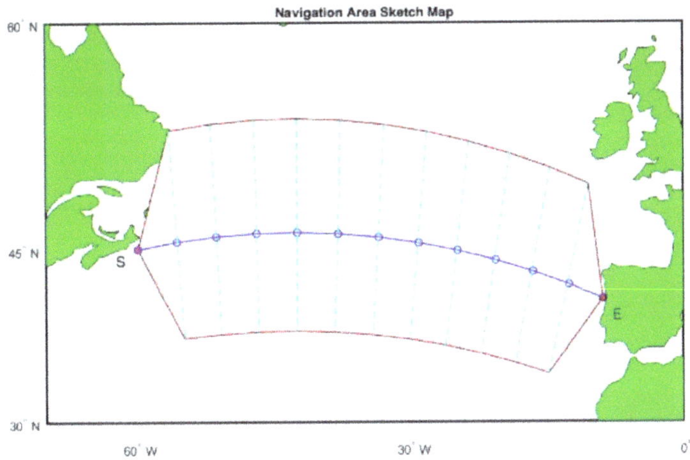

Figure 3. Expansion area. The great circle route is the reference route.

Here, Up_i and Low_i represent the position coordinates of the upper boundary point and the lower boundary point, respectively. Individuals are randomly generated according to a uniform distribution to make the initial population evenly distributed in the entire solution space and increase the diversity of the initial population according to Equation (27). After multiple generations, the initial population is obtained.

$$X' = rand \cdot \{UB - LB\} + LB. \tag{27}$$

2.5. Particle Swarm Optimization and Genetic Algorithm

PSO and GA are intelligent optimization algorithms developed recently and are widely used in path planning. The PSO algorithm is a random search algorithm based on group cooperation [33]. In PSO, the particle updates itself by tracking two "extremums," among which the first is called the individual optimal solution and the second is called the global optimal solution. PSO updates the position through Equations (28) and (29).

$$v_{id}^{k+1} = \omega \cdot v_{id}^{k} + c_1 \cdot rand \cdot \left(p_{pBest} - x_{id}^{k}\right) + c_2 \cdot rand \cdot \left(p_{gBest} - x_{id}^{k}\right), \tag{28}$$

$$x_{id}^{k+1} = x_{id}^{k} + v_{id}^{k+1}, \quad i = 1, 2, \ldots, m; \; d = 1, 2, \ldots, D, \tag{29}$$

where v_{id}^{k} is the particle velocity of the dth dimension of the mth particle in the kth iteration and x_{id}^{k} is the particle position of the dth dimension of the mth particle in the kth iteration. ω, c_1, and c_2 are the coefficients, and rand is a uniform random number in the range of $(0, 1)$.

GA is an optimization method developed based on Darwin's theory of natural selection [34]. When implementing GA, each individual is given a fitness level, which is the standard for measuring the quality of the individual. Then, the population is subjected to the selection, crossover, and mutation operations for obtaining a new population. Iterations are performed until the desired result is obtained.

2.6. Multicriteria Route Planning Algorithm

PSO and GA obtain good results when solving single-criteria problems but not when solving multicriteria problems. Therefore, this study combines PSO with GA and proposes multicriteria PSO–GA to solve the problem of ship route planning. The proposed algorithm mainly combines the particle cooperative operation associated with PSO, crossover operation, mutation operation, and multigroup elite selection operation in GA and improves the distribution of the Pareto solution set [34]. The flow chart of the algorithm is shown in Figure 4.

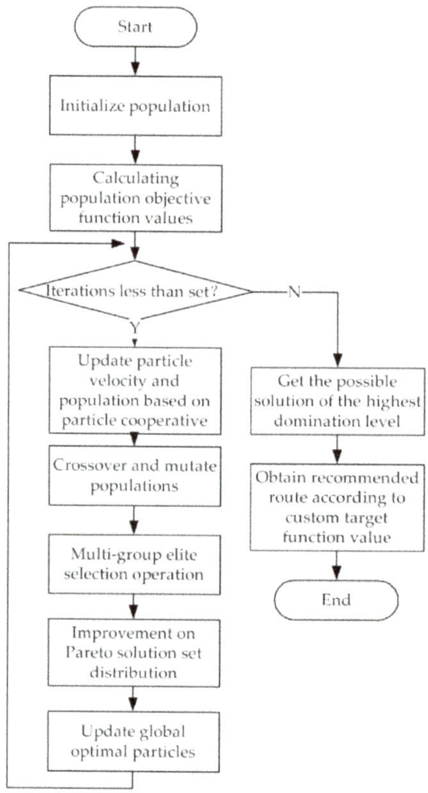

Figure 4. Flow chart of the multicriteria route planning algorithm.

The algorithm flow will be analyzed in detail below to introduce the operation steps in the algorithm more clearly.

2.6.1. Particle Cooperative Operation

The particles in the population constantly update their positions through information exchange according to the individual and global optimal solutions. At the beginning of the iteration, the ω in Equation (28) must be large to increase the diversity of the population. As the number of iterations increases, the solution tends to become optimal. ω is gradually reduced to improve the convergence speed of the algorithm. Therefore, the speed update is determined according to Equation (30), whereas the position update is determined according to Equation (29), where $MaxGen$ is the maximum number of iterations. The updated speed will be lower than $MaxV$. Further, the route through land will be regenerated.

$$v_{id}^{k+1} = \left(1 - \left(\frac{0.4}{MaxGen}\right) \cdot i\right) \cdot v_{id}^{k} + c_1 \cdot rand \cdot \left(p_{pBest} - x_{id}^{k}\right) + c_2 \cdot rand \cdot \left(p_{gBest} - x_{id}^{k}\right). \quad (30)$$

2.6.2. Crossover Operation

In this study, the arithmetic crossover method is used to generate two new individuals from the linear combination of two individuals. By assuming that the two individuals are

X_1^g and X_2^g, the calculation method of the new individual after the crossover operation is as follows:

$$\begin{cases} X_1^{g+1} = X_1^g + \alpha \cdot (X_2^g - X_1^g)' \\ X_2^{g+1} = X_2^g + \alpha \cdot (X_1^g - X_2^g)' \end{cases}, \tag{31}$$

where α is a vector parameter with the same dimension as the individual and is a random number in the $(0, 1)$ interval. When the route after the crossover operation passes over land, the crossover operation must be repeated until the required route is obtained or the set number of repetitions is reached.

2.6.3. Mutation Operation

This study uses the reference route as the mean value to perform a single-point Gaussian mutation to improve the local search capability of the algorithm and search for the optimal route near the reference route. The Gaussian distribution can be determined as follows:

$$f(x) = \frac{1}{\sqrt{2\pi}\sigma} e^{-\frac{(x-\mu)^2}{2\sigma^2}}, \tag{32}$$

where μ is the coordinate value of the mutation position. The variable generated according to the Gaussian distribution has a 99.73% probability of falling in the interval $(\mu - 3\sigma, \mu + 3\sigma)$. Therefore, σ can be determined as follows:

$$\sigma = \frac{|UB_i - X_i|}{3}. \tag{33}$$

2.6.4. Multigroup Elite Selection Operation

This study comprehensively processes the original population, the population after particle coordination, and the population after crossover and mutation to improve the quality of the global optimal solution. The three populations are merged to perform nondominated sorting, and the individual in the front after sorting is considered to be the new individual of the next generation. Domination and nondomination can be given as follows [34]:

$$u = F(p') = min\{f_1(p'), f_2(p'), \ldots, f_{n_0}(p')\}, \tag{34}$$

$$u' = F(q') = min\{f_1(q'), f_2(q'), \ldots, f_{n_0}(q')\}, \tag{35}$$

$$ff[\forall i \in \{1, \ldots, n_0\}, u_i \leq u_i'] \wedge [\exists i \in \{1, \ldots, n_0\}, u_i < u_i'], \tag{36}$$

where p' and q' are the decision variable vectors and the position information of the two particles is represented in the algorithm. u and u' are the optimization target vectors of the two particles p' and q', respectively. Because there are three optimization targets in this study, $n = 3$. If the performance vectors u and u' satisfy Equation (36), then particle p' dominates q'. If a particle neither dominates nor is dominated by other particles, the particle is called a nondominated solution. The set of all particles that satisfy the nondominated solution is called a nondominated solution set.

The crowding distance between the particles at each domination level can be defined as follows to further evaluate the performance of the particles in each domination level.

$$Crowd(m) = \sum_{j=1}^{n_0} \frac{f_j(m+1) - f_j(m-1)}{f_{jMax} - f_{jMin}}, \quad m = 2, 3, \ldots, N' - 1, \tag{37}$$

where $Crowd(m)$ is the crowded distance of the mth particle, f_j is the jth objective function value, f_{jMax} and f_{jMin} are the maximum and minimum values of the jth optimization target, respectively, N' is the number of particles at the same dominance level, and the crowding distance for edge particles is set to infinity.

Finally, after completing the improvement of the Pareto solution set in Section 2.6.5, a particle is randomly selected as the global optimal particle from among the particles with the highest dominating level and a degree of crowding not equal to infinity. Moreover, in each iteration, the dominant status of individuals in the population will be calculated according to Equation (36), and the population will be sorted according to the dominant level and crowding degree. Individuals with higher dominance levels will be preferentially selected to the next generation, the higher the dominance level of an individual, the closer it will be to the Pareto optimal frontier. After many iterations, individuals in the population will continue to move closer to the Pareto optimal frontier [34].

2.6.5. Improved Pareto Solution Set Distribution

In the Pareto optimal solution set obtained in Section 2.6.4, there may be cases in which multiple solutions are clustered in adjacent areas. This study will use the following steps to improve the Pareto solution distribution generated in Section 2.6.4 to make the Pareto solution set more evenly distributed in the entire solution space and avoid local convergence of the algorithm.

(1) Sort the individuals in the solution set with the highest dominance level according to the navigation risk value from small to large.

(2) Calculate the Euclidean distance, dis_i, between the nondominated solution i and the nondominated solution $i+1$ in the target space. Here, $f_i(j)$ represents the jth objective function value of the ith nondominated solution.

$$dis_i = \sqrt{\sum_{j=1}^{n}[f_i(j) - f_{i+1}(j)]^2}, \tag{38}$$

(3) Determine whether dis_i is less than the specified value dis_set, which is determined according to Equation (39), where num represents the number of nondominated solutions. If $dis_{i-1} < dis\_set$ and $dis_i < dis\_set$, delete the nondominated solution i and calculate the Euclidean distance between the nondominated solution $i-1$ and the nondominated solution $i+1$ and assign it to dis_{i-1}.

$$dis\_set = \frac{\sum_{i=1}^{num-1} dis_i}{n-1}, \tag{39}$$

(4) Determine whether the number of nondominated solutions exceeds the set value $SetNum$. If yes, move to step 5; otherwise, skip to step 6.

(5) Delete $p = num - SetNum$ nondominated solutions in the crowded area to simplify the nondominated solutions in the crowded area. The flowchart is shown in Figure 5.

$$dis = \{dis_1, dis_2, \cdots, dis_{num-2}, dis_{num-1}\}. \tag{40}$$

(6) Return an improved set of Pareto solutions.

2.6.6. Recommended Routes

According to the Pareto frontier and Pareto solution set obtained using the proposed algorithm, the recommended route can be given as follows:

$$Z = \min\left(\left(\sum_{j=1}^{n} a_j \cdot |c_{ij} - y_j|^2\right)^{\frac{1}{2}}\right), \quad \sum_{j=1}^{n} a_j = 1, \quad (i = 1, 2, \ldots, M), \tag{41}$$

where c_{ij} is the normalized value of the jth objective function of the ith particle, y_j is the normalized value of the expected objective function, which, respectively, indicate the navigation time, fuel consumption, and acceptable navigation risk set by the shipping company or captain before the start of the navigation mission. M is the number of particles in the Pareto solution set, and Z is the recommended route in which the conditions given on the right side of the equation are satisfied.

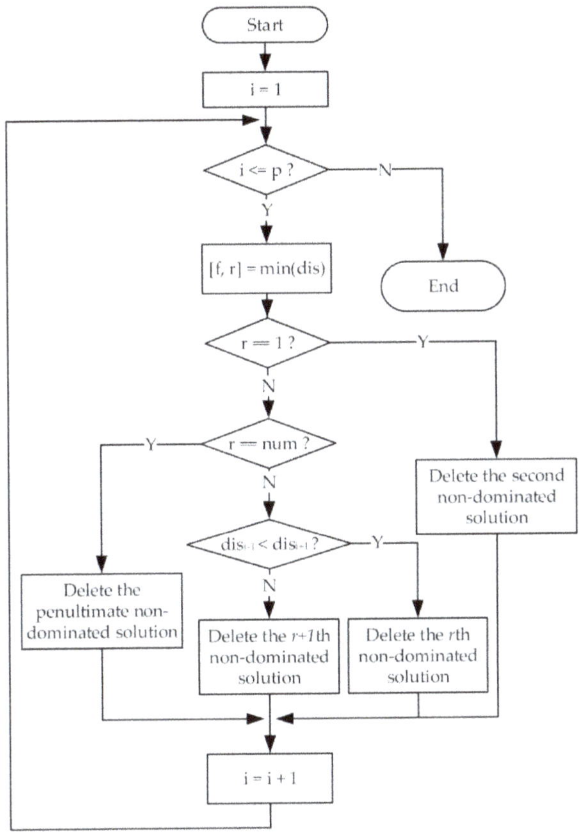

Figure 5. Flow chart of step 5 for improved Pareto distribution.

3. Results

3.1. Algorithm Parameter Setting

The experimental ship in this study was an S-175 container ship. The ship parameters are shown in Table 2.

Table 2. S-175 container ship parameters.

Parameter	Length	Breadth	Depth	Draught	Displacement	Metacentric Height
Value	175 m	25.4 m	15.4 m	9.5 m	23,740 m^3	1.022 m

The sailing area expands in both the directions along the angle bisector of the reference route. In experiment one and experiment two, the extension was 8 degrees in two directions, respectively. Meteorological data comes from the public database of the European Center for Mid-range Weather Forecast. Some ship speed and fuel consumption data used for curve fitting are shown in Table 3. In Equation (21), a = 0.1255 (95% confidence interval: (0.1171, 0.1339)), b = 0.2444 (95% confidence interval: (0.2395, 0.2493)). Root mean squared error (RMSE): 0.02781. Coefficient of determination (R-square): 0.9995. The parameters of the multicriteria algorithm are presented in Table 4.

In this table, *Gen* represents the number of algorithm iterations, *PoPu* represents the number of populations,c_1 and c_2 represent the parameters in PSO, *Mu* represents the uniform mutation probability, *Cr* represents the crossover probability, *SetNum* represents the specified maximum number of nondominated solutions, and *MaxV* represents the maximum velocity of the particle swarm. For Equation (41), in the three experiments, the weight coefficients a_j of risk, fuel consumption, and time are set to 0.4, 0.3, and 0.3, respectively. This study establishes a visual simulation interface for the convenience of ship route optimization, as shown in Figure 6.

Table 3. Ship speed and fuel consumption data.

Velocity (kn)	**10**	**10.5**	**11**	**11.5**	**12**	**12.5**	**13**	**13.5**	**14**	**14.5**	**15**
Fuel consumption (tons/hour)	1.40	1.62	1.86	2.11	2.38	2.68	2.30	3.36	3.85	4.34	4.91

Table 4. Algorithm parameter values.

Gen	*PoPu*	c_1	c_2	*Mu*	*Cr*	*SetNum*	*MaxV*
100	50	2	2	0.2	0.7	40	0.8

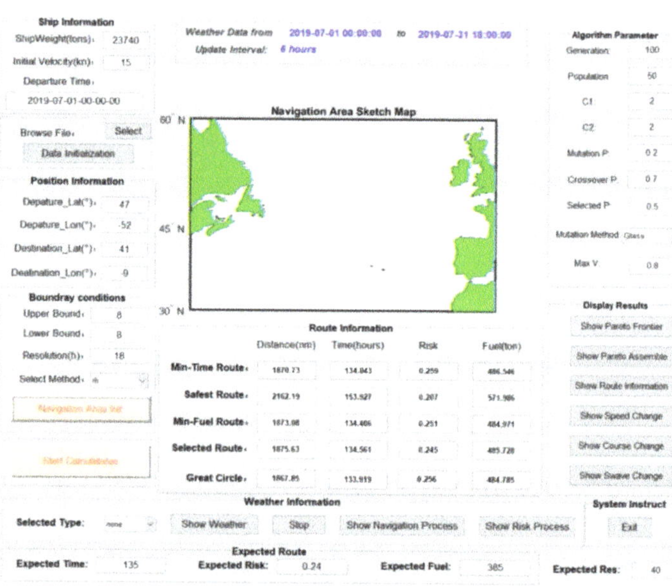

Figure 6. Multicriteria route planning simulation interface.

3.2. Experimental Results and Analysis

The multicriteria ship route of the S-175 container ship was optimized according to the algorithm parameters set in the previous subsection. This study included three experiments. The first was to plan routes in sea under good weather conditions, the second was to plan routes in sea under severe weather conditions, and the third was to plan routes under complex offshore conditions. In Experiment one and Experiment two, the number of route segments was set to 7, and the number of route segments in Experiment three was set to 10.

3.2.1. Good Weather Conditions

The starting and ending ports of the ship are St. John's (47 N, 52 W) and Porto (41 N, 9 W), respectively. According to the weather forecast, from 1 to 7 July 2019, the ships sailing along the great circle route will not encounter strong winds and waves. Under this condition, the great circle route between the two ports is set as the reference route. We conducted 20 repeated experiments for this sea state and got the average navigation time of the recommended route to be 135.850 and the variance to be 0.942. The average risk was 0.239, and the variance was 2.41×10^{-5}. The average fuel consumption was 490.533, and the variance was 48.703. A group of experiments were then selected for a detailed description.

After algorithm optimization, the Pareto optimal solution set of the multicriteria route is shown in Figure 7a. The Pareto optimal frontier solution is shown in Figure 7b. According to the route position information in Pareto solution collection, routes with the minimum navigation time, risk, and fuel consumption were obtained. The expected navigation time was set to 135 h, the navigation risk was set to 0.24, and the fuel consumption was set to 485 tons. According to the above-expected values and Equation (41), the recommended route was obtained, which is a compromise route obtained by balancing the three objectives. This route not only ensures that the navigation risk is within an acceptable range, but also reduces the navigation time and fuel consumption as much as possible. The schematic of the above four routes and the great circle route is shown in Figure 7c. The recommended route is shown in Figure 7d.

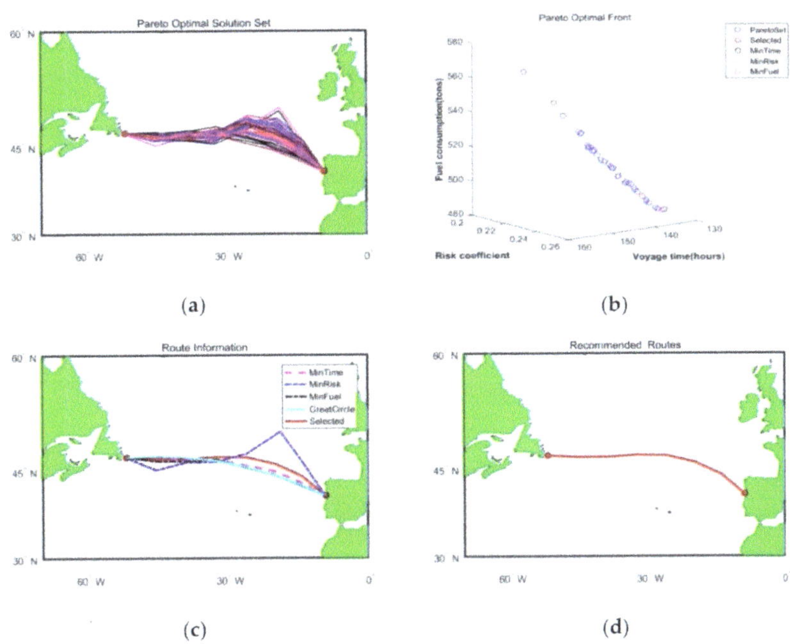

Figure 7. Ship route optimization results under good sea conditions: (**a**) the Pareto optimal solution set of the multicriteria route; (**b**) the Pareto optimal frontier solution; (**c**) great circle route and four optimized routes; and (**d**) recommended routes.

According to Table 5 and Figure 7, the minimum-time route and the minimum-fuel-consumption route basically coincide with the great circle route, and the voyage difference was only 4.5 and 9.52 nmi respectively, indicating that the algorithm has better route search performance. The minimum-time route, the safest route, and the minimum-fuel-consumption route are calculated by the Pareto solution set according to a single criterion. In the recommended route, these three factors are integrated to obtain a compromised and improved ship route. Further, the captain can obtain another recommended route according to the different expected objective function values. When sailing on the recommended route, the position of the ship at four different navigation times is shown in Figure 8. Under good sea conditions, the ship can sail along the recommended route with low navigation risk, fuel consumption, and navigation time.

Table 5. Objective function values of five routes in good sea conditions.

-	Distance (nmi)	Time (h)	Risk	Fuel (tons)
Minimum-time Route	1872.32	134.222	0.250	486.112
Safest Route	2166.99	154.289	0.209	572.946
Minimum-Fuel Route	1877.34	134.724	0.250	485.937
Great Circle Route	1867.82	133.919	0.259	484.787
Recommended Route	1867.85	133.919	0.256	484.785

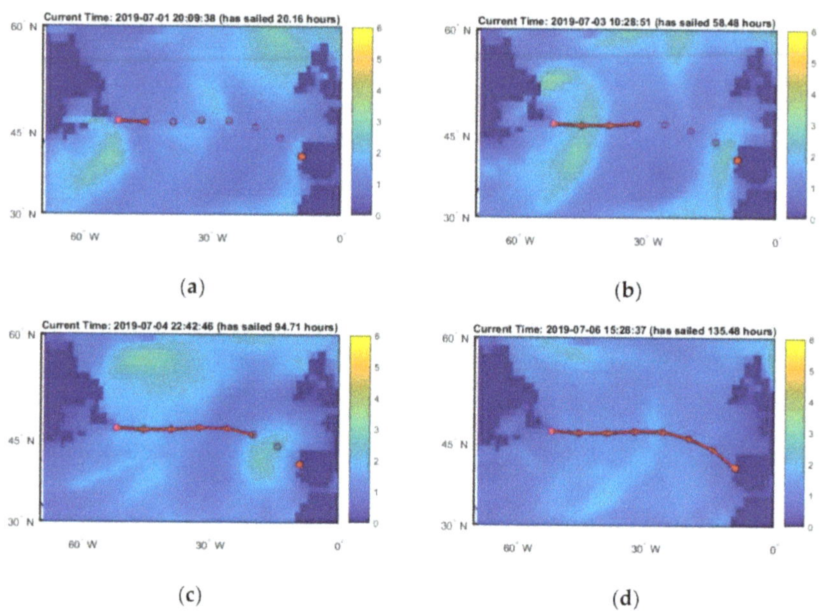

Figure 8. Position of the ship under good sea conditions at four different navigation times. (**a**–**d**) show the positions of the ship with respect to the wave height map at different times.

3.2.2. Severe Weather Conditions

A multicriteria ship route planning was performed for severe weather conditions in the ship sailing area to prove the effectiveness of the algorithm presented in this study. The starting and ending ports of the ship were St. John's (47 N, 52 W) and Porto (41 N, 9 W), respectively. According to weather forecasts, from 22 to 29 July 2019, if the ship sailed along the rhumb line, it would encounter strong winds and waves. Under such sea conditions, the rhumb line between the two ports is set as the reference route. We conducted 20 repeated experiments for this sea state and got the average navigation time of the recommended route to be 157.170 and the variance to be 0.421. The average risk was 0.416, and the variance was 2.06×10^{-4}. The average fuel consumption was 437.988, and the variance was 33.739. A group of experiments were then selected for a detailed description.

After algorithm optimization, the Pareto optimal solution set of the weather route is shown in Figure 9a. The Pareto optimal frontier solution is presented in Figure 9b. According to the route position information in the Pareto solution, routes with the minimum navigation time, safety risk, and fuel consumption can be obtained. The expected navigation time was set to 152 h, the navigation risk was set to 0.42, and the fuel consumption was set to 430 tons. The five optimized routes are shown in Figure 9c. The recommended route is shown in Figure 9d, and the specific data of the optimized route are presented in Table 6.

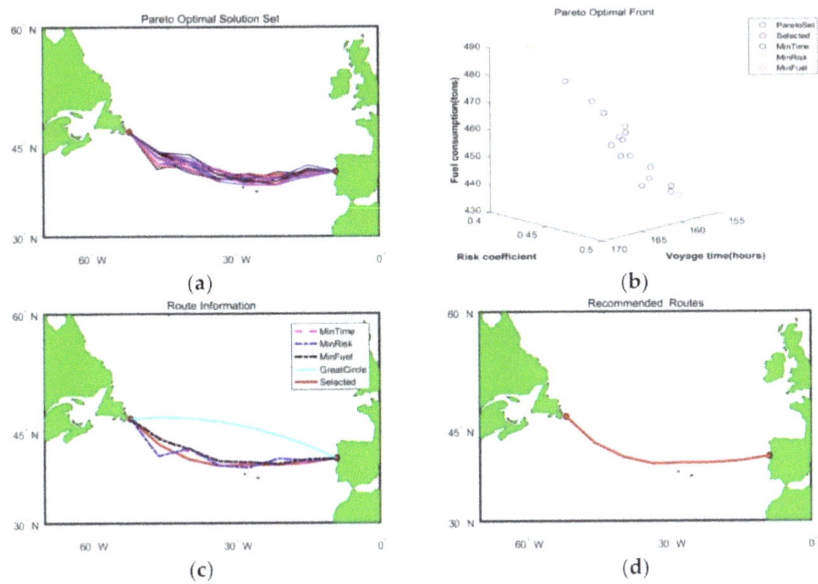

Figure 9. Ship route optimization results under severe conditions: (**a**) the Pareto optimal solution set of the multicriteria route; (**b**) the Pareto optimal frontier solution; (**c**) great circle route and four optimized routes; and (**d**) recommended routes.

Table 6. Objective function values of five routes in severe conditions.

-	Distance (nmi)	Time (h)	Risk	Fuel (tons)
Minimum-time Route	2027.22	156.785	0.469	437.612
Safest Route	2179.30	165.677	0.405	487.839
Minimum-fuel Route	2024.04	156.922	0.476	435.241
Great Circle Route	1867.85	155.732	0.562	366.094
Recommended Route	2075.98	158.815	0.442	458.005

According to Table 6 and Figure 9, when the ship sails along the great circle route, the voyage was reduced by 159.37 and 208.13 nmi than the minimum navigation time route and recommended route, respectively. However, the navigation time was only reduced by 1.118 and 3.083 h, respectively. Thus, when the ship is sailing along the great circle route, large wind waves cause serious speed loss to the ship and the ship is exposed to great risks. Therefore, sailing along the great circle route is not suitable. According to Equation (41) and the above-mentioned expected objective function value, a compromise route with an objective function value close to the expected value was selected in the Pareto solution set. When sailing according to the recommended route, the position of the ship at four different navigation times is shown in Figure 10. Under severe sea conditions, the ship could sail along the recommended route with low navigation risk, fuel consumption, and navigation time.

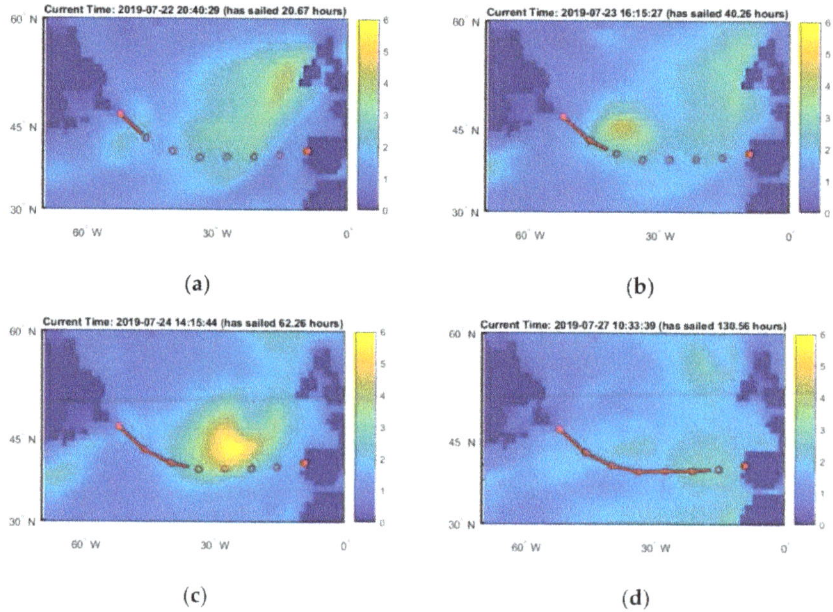

Figure 10. Position of the ship in severe conditions at four different navigation times. (**a**–**d**) show the position of the ship with respect to the wave height map at different times.

3.2.3. Complex Offshore Conditions

Multicriteria ship route planning was performed in the waters near the Cape of Good Hope to verify the effectiveness of the algorithm for route planning near coastal waters. The starting position of the ship was (23 S, 6 E), and the destination position was (26 S, 42 E). According to the weather forecast, from 1st to 7th March 2019, the ships sailing along the reference route will encounter heavy winds and waves. We conducted 20 repeated experiments for this sea state and got the average navigation time of the recommended route to be 183.668 and the variance to be 2.725. The average risk was 0.404, and the variance was 4.05×10^{-5}. The average fuel consumption was 526.396, and the variance was 29.624. A group of experiments were then selected for a detailed description.

In this sea state, the Pareto optimal solution set of the ship route after algorithm optimization is shown in Figure 11a. The Pareto optimal frontier solution is shown in Figure 11b. According to the route position information in the Pareto solution set, the smallest-sailing-time route, the least-risk route, and the least-fuel-consumption route can be separately calculated. The expected navigation time was set to 180 h, the navigation risk was set to 0.4, and the fuel consumption was set to 510 tons for obtaining the recommended route. The four routes are presented in Figure 11c. The recommended route is shown in Figure 11d, and specific data related to the four optimized routes are shown in Table 7.

Table 7. Objective function values of five routes in complex offshore conditions.

-	Distance (nmi)	Time (h)	Risk	Fuel (tons)
Minimum-time Route	2334.73	182.562	0.412	526.436
Safest Route	2520.17	196.435	0.402	573.419
Minimum-fuel Route	2344.30	184.569	0.420	522.402
Recommended Route	2335.18	182.754	0.413	525.953

According to Table 7 and Figure 11, routes satisfying different criteria under comprehensive offshore sea conditions can be obtained after algorithm optimization. The recommended route is based on the expected objective function value. According to Equation (41), a recommended route close to the expected target value was selected in the Pareto optimal solution set. Compared with the minimum-time route, the recommended route had 0.01 higher navigation risk and 0.192 h longer navigation time, but it saved 0.483 tons of fuel. The position of the ship when sailing on the recommended route at four different navigation times is shown in Figure 12. Under complex sea conditions, it was possible to sail along the recommended route with low navigation risk, fuel consumption, and navigation time.

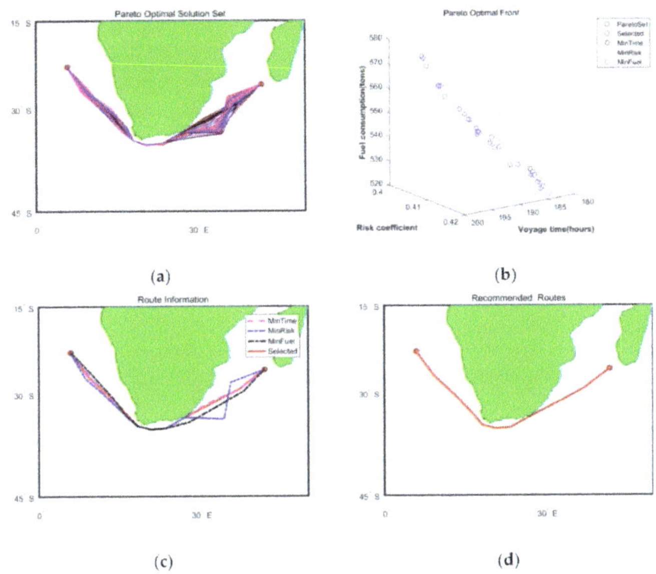

Figure 11. Ship route optimization results under complex offshore conditions: (**a**) the Pareto optimal solution set of the multicriteria route; (**b**) the Pareto optimal frontier solution; (**c**) great circle route and four optimized routes; and (**d**) recommended routes.

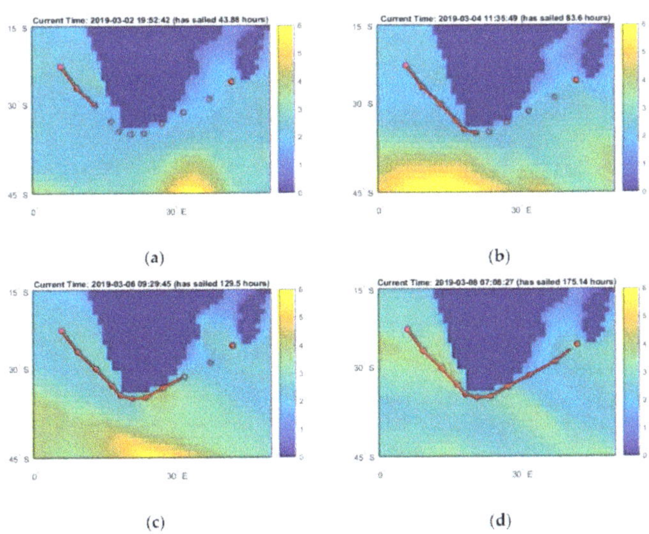

Figure 12. Position of the ship under complex offshore conditions at four different navigation times. (**a**–**d**) show the position of the ship with respect to the wave height map at different times.

4. Discussion

According to the characteristics of ship navigation, this study establishes a multicriteria route planning framework for ships comprising optimization criteria, ship speed analysis, model construction, multicriteria algorithms, route evaluation, and route selection. There are three criteria for ship route optimization: minimum navigation time, minimum navigation risk, and minimum fuel consumption. For navigation risk, this study proposes a comprehensive route risk assessment model that considers wind, waves, and seakeeping. Regarding fuel consumption, this study established a mathematical model of ship fuel consumption through fitting methods. Based on this framework, this study proposes a ship multicriteria route planning method using the improved PSO–GA. The proposed algorithm not only has the fast convergence of the particle swarm algorithm but also improves the diversity of solutions by applying the crossover operation, selection operation, and multigroup elite selection operation of the genetic algorithm and improving the Pareto optimal frontier distribution. By solving the route planning algorithm, the routes with the minimum navigation time, fuel consumption, and navigation risk as well as the recommended route can be obtained. In addition, the captain can customize the appropriate route according to the route solution set to ensure that the ship can sail efficiently, safely, and economically. We conduct experiments for three different situations to verify the performance of the algorithm. According to the experimental results, the recommended route can be obtained by integrating the three aforementioned indicators, based on which the area with severe winds and waves can be avoided and minimum navigation time and fuel consumption can be realized. Therefore, the ship multicriteria route planning algorithm proposed in this study is feasible and effective.

Although this study has established a multicriteria route planning algorithm for ships, there are still shortcomings, which must be improved and perfected, mainly as follows.

- Risk assessment module: In this study, only the impact of winds, waves, and seakeeping on the ship is considered for risk assessment, whereas visibility and the impact of other ships at sea on the ship are ignored. Future studies will consider more ship risk factors.
- Recommended route strategy: In this study, the recommended route can be accurately determined based on the three expected target values. However, it is difficult to determine the target value under the unknown conditions associated with the historical route.

Author Contributions: Conceptualization, W.Z. and H.W.; methodology, W.Z.; software, W.Z.; validation, W.Z., Y.W., Z.Z. and H.W.; formal analysis, W.Z.; investigation, W.Z.; resources, W.Z.; data curation, Y.W. and Z.Z.; writing—original draft preparation, W.Z.; writing—review and editing, W.Z.; visualization, W.Z.; supervision, H.W.; project administration, H.W.; funding acquisition, H.W. All authors have read and agreed to the published version of the manuscript.

Funding: This study was funded by the Russian Foundation for Basic Research (RFBR) (no. 20-07-00531).

Institutional Review Board Statement: Not applicable.

Informed Consent Statement: Not applicable.

Data Availability Statement: Not applicable.

Acknowledgments: The authors are grateful to the anonymous reviewers for their valuable comments and suggestions that helped improve the quality of this manuscript.

Conflicts of Interest: The authors declare no conflict of interest.

References

1. Chircop, A. The IMO Initial Strategy for the Reduction of GHGs from International Shipping: A Commentray. *Int. J. Mar. Coast. Law* **2019**, *34*, 482–512. [CrossRef]
2. Wang, H.; Mao, W.; Erilksson, L. Efficiency of a Voluntary Speed Reduction Algorithm for a Ship's Great Circle Sailing. *Transnav. Int. J. Mar. Navig. Saf. Sea Transp.* **2020**, *14*, 301–308. [CrossRef]

3. James, R.W. Application of wave forecasts to marine navigation. *Comp. Biochem. Physiol. A Comp. Physiol.* **1957**, *43*, 195–205.
4. Hagiwara, H.; Spans, J.A. Practical weather routing of sail-assisted motor vessels. *J. Navig.* **1987**, *40*, 96–119. [CrossRef]
5. Lin, Y.H.; Fang, M.C.; Yeung, R.W. The optimization of ship weather-routing algorithm based on the composite influence of multi-dynamic elements. *Appl. Ocean Res.* **2013**, *43*, 184–194. [CrossRef]
6. Smierzchalski, R.; Szlapczynska, J. Adopted isochrone method improving ship safety in weather routing with evolutionary approach. *Int. J. Reliab. Qual. Saf. Eng.* **2007**, *14*, 635–645.
7. Shao, W.; Zhou, P.; Thong, S.K. Development of a novel forward dynamic programming method for weather routing. *J. Mar. Sci. Technol.* **2012**, *17*, 239–251. [CrossRef]
8. Sen, D.; Padhy, C.P. An approach for development of a ship routing algorithm for application in the North Indian Ocean region. *Appl. Ocean Res.* **2015**, *50*, 173–191. [CrossRef]
9. Mannarini, G.; Pinardi, N.; Coppini, G.; Oddo, P.; Iafrati, A. VISIR-I: Small vessels—Least-time nautical routes using wave forecasts. *Geosci. Model Dev. Discuss.* **2016**, *8*, 7911–7981. [CrossRef]
10. Wang, H.B.; Li, X.G.; Li, P.F.; Veremey, E.I.; Sotnikova, M.V. Application of Real-Coded Genetic Algorithm in Ship Weather Routing. *J. Navig.* **2018**, *71*, 989–1010. [CrossRef]
11. Chuang, T.N.; Lin, C.T.; Kung, J.Y.; Lin, M.D. Planning the route of container ships: A fuzzy genetic approach. *Expert Syst. Appl.* **2010**, *37*, 1948–1956. [CrossRef]
12. Wang, L.P.; Zhang, Z.; Zhu, Q.D.; Ma, S. Ship Route Planning Based on Double-Cycling Genetic Algorithm Considering Ship Maneuverability Constraint. *IEEE Access* **2020**, *8*, 190746–190759. [CrossRef]
13. Vlachos, D.S. Optimal Ship Routing Based on Wind and Wave Forecasts. *Appl. Numer. Anal. Comput. Math.* **2004**, *1*, 547–551. [CrossRef]
14. Tsou, M.C.; Cheng, H.C. An Ant Colony Algorithm for efficient ship routing. *Pol. Marit. Res.* **2013**, *20*, 28–38. [CrossRef]
15. Zhang, G.Y.; Wang, H.B.; Zhao, W.; Guan, Z.Y.; Li, P.F. Application of Improved Multi-objective Ant Colony Optimization Algorithm in Ship Weather Routing. *J. Ocean Univ. China* **2021**, *20*, 45–55. [CrossRef]
16. Vettor, R.; Guedes Soares, C. Detection and analysis of the main routes of voluntary observing ships in the North Atlantic. *J. Navig.* **2015**, *68*, 397–410. [CrossRef]
17. Vettor, R.; Guedes Soares, C. Development of a ship weather routing system. *Ocean Eng.* **2016**, *123*, 1–14. [CrossRef]
18. He, Y.K.; Zhang, D.; Zhang, J.F.; Zhang, M.Y.; Li, T.W. Ship Route Planning Using Historical Trajectories Derived from AIS Data. *Transnav. Int. J. Mar. Navig. Saf. Sea Transp.* **2019**, *13*, 69–76. [CrossRef]
19. Abd-El-Wahed, W.F.; Mousa, A.A.; El-Shorbagy, M.A. Integrating particle swarm optimization with genetic algorithms for solving nonlinear optimization problems. *J. Comput. Appl. Math.* **2011**, *235*, 1446–1453. [CrossRef]
20. Liu, Z.; Liu, J.X.; Zhou, F.; Liu, R.W.; Xiong, N.X. A Robust GA/PSO-Hybrid Algorithm in Intelligent Shipping Route Planning Systems for Maritime Traffic Networks. *J. Internet Technol.* **2018**, *19*, 1635–1644.
21. IMO. Revised Guidance to the Master for Avoiding Dangerous Situations in Adverse Weather and Sea Conditions: MSC.1/Circ.1228. 2007. Available online: https://wwwcdn.imo.org/localresources/en/OurWork/Safety/Documents/Stability/MSC.1-CIRC.1228.pdf (accessed on 15 January 2021).
22. IMO. Adoption of the International Code on Intact Stability, 2008 (2008 Is Code). Available online: https://wwwcdn.imo.org/localresources/en/KnowledgeCentre/IndexofIMOResolutions/MSCResolutions/MSC.267(85).pdf (accessed on 15 January 2021).
23. Sheng, Z.B.; Liu, Y.Z. *Principles of Ships (Volume 2)*, 13th ed.; Shanghai Jiaotong University Press: Shanghai, China, 2017; pp. 64–166.
24. Lewis, E.V. *Principles of Naval Architecture*, 2nd ed.; The Society of Naval Architects and Marine Engineers: Jersey City, NJ, USA, 1989; Volume 3.
25. Faltinsen, O.M. *Sea Loads on Ships and Offshore Structures*; Cambridge University Press: Cambridge, UK, 1990.
26. Stevens, S.C.; Parsons, M.G. Effects of Motion at Sea on Crew Performance: A Survey. *Mar. Technol. SNAME News* **2002**, *39*, 29–47. [CrossRef]
27. Pipchenko, O.D.; Zhukov, D.S. Ship Control Optimization in Heavy Weather Conditions. In Proceedings of the 2010 11th AGA IAMU, Busan, Korea, 16–18 October 2010.
28. Zhang, W.; Zou, Z.J. Time domain simulations of the wave-induced motions of ships in maneuvering condition. *J. Mar. Sci. Technol.* **2016**, *21*, 154–166. [CrossRef]
29. Pennino, S.; Gaglione, S.; Innac, A.; Piscopo, V.; Scamardella, A. Development of a New Ship Adaptive Weather Routing Model Based on Seakeeping Analysis and Optimization. *J. Mar. Sci. Eng.* **2020**, *8*, 270. [CrossRef]
30. Von Westarp, A.G. A new model for the calculation of the bunker fuel speed-consumption relation. *Ocean Eng.* **2020**, *204*, 107262. [CrossRef]
31. Kim, M.; Hizir, O.; Turan, O.; Day, S.; Incecik, A. Estimation of added resistance and ship speed loss in a seaway. *Ocean Eng.* **2017**, *141*, 465–476. [CrossRef]
32. Liu, F. Research on the problem of ship stalling in wind and waves. *J. Dalian Marit. Univ.* **1992**, *4*, 347–351.
33. Kennedy, J.; Eberhart, R. Particle Swarm Optimization. In Proceedings of the IEEE International Conference on Neural Networks, Perth, WA, Australia, 27 November–1 December 1995; Volume 4, pp. 1942–1948.
34. Deb, K.; Pratap, A.; Agarwal, S.; Meyarivan, T. A fast and elitist multiobjective genetic algorithm: NSGA-II. *IEEE Trans. Evol. Comput.* **2002**, *6*, 182–197. [CrossRef]

Journal of
Marine Science
and Engineering

MDPI

Article

3-Dimensional Modeling and Attitude Control of Multi-Joint Autonomous Underwater Vehicles

Lin Yu [1], Qinghao Meng [1] and Hongwei Zhang [2,*]

[1] School of Electrical and Information Engineering, Tianjin University, Tianjin 300072, China; 2018234235@tju.edu.cn (L.Y.); qh_meng@tju.edu.cn (Q.M.)
[2] School of Mechanical Engineering, Tianjin University, Tianjin 300072, China
* Correspondence: zhanghongwei@tju.edu.cn; Tel.: +86-13821321024

Abstract: To achieve rapid and flexible vertical profile exploration of deep-sea hybrid structures, a multi-joint autonomous underwater vehicle (MJ-AUV) with orthogonal joints was designed. This paper focuses on the 3-dimensional (3D) modeling and attitude control of the designed vehicle. Considering the situation of gravity and buoyancy imbalance, a 3D model of the MJ-AUV was established according to Newton's second law and torque balance principle. And then the numerical simulation was carried out to verify the credibility of the model. To solve the problems that the pitch and yaw attitude of the MJ-AUV are coupled and the disturbance is unknown, a linear quadratic regulator (LQR) decoupling control method based on a linear extended state observer (LESO) was proposed. The system was decoupled into pitch and yaw subsystems, treated the internal forces and external disturbances of each subsystem as total disturbances, and estimated the total disturbances with LESO. The control law was divided into two parts. The first part was the total disturbance compensator, while the second part was the linear state feedback controller. The simulation results show that the overshoot of the controlled system in the dynamic process is nearly 0 rad, reaching the design value very smoothly. Moreover, when the controlled system is in a stable state, the control precision is within 0.005%.

Keywords: multi-joint autonomous underwater vehicle (MJ-AUV); 3-dimensional modeling; LQR; LESO

Citation: Yu, L.; Meng, Q.; Zhang, H. 3-Dimensional Modeling and Attitude Control of Multi-Joint Autonomous Underwater Vehicles. *J. Mar. Sci. Eng.* **2021**, *9*, 307. https://doi.org/10.3390/jmse9030307

Academic Editors: Rosemary Norman and Evgeny Veremey

Received: 14 January 2021
Accepted: 8 March 2021
Published: 10 March 2021

1. Introduction

Ocean exploration technology is one of the difficult problems in frontier science and engineering in the ocean field. The deep-sea region hybrid is an important factor in maintaining global energy balance and driving deep ocean circulation. Therefore, it is of great strategic significance to use advanced technology to explore the deep-sea region hybrid structures [1]. The complex structure of the seabed, as well as unknown extreme fluid systems such as cold springs and hydrothermal fluids, make the work of exploration more difficult.

Most of the existing deep-sea submersibles struggle to meet the capabilities necessary to explore deep-sea hybrid structures rapidly. The glider [2,3] achieves pose control by adjusting the remaining buoyancy and the position of the mass center, but its motion trajectory is jagged and single, speed is slow, and maneuverability is poor. Most autonomous underwater vehicles (AUVs) [4] are rigid single-body structures and use tail rudder and attitude adjustment systems to control the movement direction. In order to further improve the ability of flexible steering of the AUV to allow rapid observation of the deep-sea 3-dimensional (3D) environment, a multi-joint AUV (MJ-AUV) was designed. It consists of three parts in series: diversion cabin, navigation/control cabin, and propulsion cabin, each of which is connected by a pair of orthogonal joints. In addition, its tail is equipped with a propeller. To adjust the yaw attitude and pitch attitude of the body, the MJ-AUV can change the hydrodynamic appearance of the vehicle by rotating the joints.

The MJ-AUV is a multi-rigid-body rootless system with high nonlinear and strong coupling characteristics, the kinematic and dynamic modeling of which is the basis of studying its motion behavior characteristics and control problems. Xia et al. [5] established a horizontal dynamic model of a fish-like robot based on Kane's method. Aiming at the structure of the underwater gliding snake-like robot, Tang et al. [6] built a gliding and serpentine swimming model based on the momentum theorem, moment of momentum theorem, and recursive Newton–Euler method. Based on the principle of force and moment balance, Kelasidi et al. [7–9] built a horizontal plane and slope dynamics model of the underwater snake robot in the inertial system. In addition, the Euler–Lagrange method [10] and the Schiehlen method [11] have also been used to deal with multi-rigid body modeling.

The controller design is the key technology to enable underwater vehicles to complete a deep-sea exploration mission. Professor Pettersen's team from the University of Norway has made many contributions in the field of underwater multi-joint robot control, such as planar path tracking in the presence of ocean currents [12] and integral line-of-sight guidance for path following control [13]. Fischer et al. [14] propose an error controller using continuous robust integration to compensate for the uncertainty of the AUV model and have carried out experiments under controlled and open-water environments to verify the effectiveness of the controller. Zhao et al. [15] propose an adaptive plus disturbance observer (DOB) controller for depth and attitude control of the AUV. The controller consists of DOB as an inner-loop compensator and a non-regressor based adaptive controller as an outer-loop controller. In addition, the experiments verify that the controller has strong robustness. Wu et al. [16], Wang et al. [17] and Kang et al. [18] make improvements on the basis of the adaptive controller, which is verified in the field of motion control of AUVs. Zhang et al. [19] present a sliding mode variable structure control algorithm, and simulation results show that this algorithm has advantages of high control accuracy and strong robustness. Rodriguez et al. [20] combine sliding mode control with adaptive control and propose a sliding mode adaptive controller, which is compared with non-adaptive control and PD control to verify the effectiveness of the controller. References [21–25] have made improvements based on the active disturbance rejection controller, combining sliding mode controller, self-searching optimal algorithm, or other methods, to improve the accuracy and anti-interference performance of the AUV motion control. In addition to the above methods, in recent years, scholars are also studying the application of reinforcement learning [26,27] and artificial intelligence algorithms [28] in the field of AUV control.

MJ-AUV is a complex system with high nonlinear, strong coupling, large time delay, and unknown disturbance, and establishing an accurate mathematical model for the MJ-AUV is difficult. In the course of pitching and yaw attitude control, the MJ-AUV belongs to a typical multi-input and multi-output system, with the variation of two joint angles as the input and the pitch and yaw angles of the body as the output, making the controller design more difficult. Hence, the main contributions of this paper are as follows:

(1) A novel AUV with orthogonal joints was proposed and designed for rapid and flexible vertical profile exploration of deep-sea hybrid structures, and the 3D motion model of the designed AUV was established according to Newton's second law and the principle of moment balance.
(2) To reduce the coupling degree of the controller and improve the accuracy of attitude control, a linear quadratic regulation (LQR) decoupling control method based on a linear extended state observer (LESO) [29] was proposed.

The remainder of this paper is organized as follows. Section 2 establishes the 3D motion model of the MJ-AUV. The LQR decoupling control method based on LESO is presented in Section 3. In Section 4, the pitch and yaw control of the MJ-AUV is simulated on the MATLAB/SIMULINK platform, followed by the conclusion in Section 5.

2. Modelling

This section introduces the structure design, kinematics, and dynamics analysis of the MJ-AUV and presents the 3D motion model.

2.1. Structure of the MJ-AUV

As shown in Figure 1, from front to back, the MJ-AUV is mainly composed of the diversion cabin, navigation/control cabin, and propulsion cabin, which are connected by two orthogonal (pitch and yaw degrees of freedom) joints. A propeller and a fixed rudder are installed at the tail to enhance the body stability. The sensors such as hydrophones and thermohaline depth meter can be mounted according to specific task requirements.

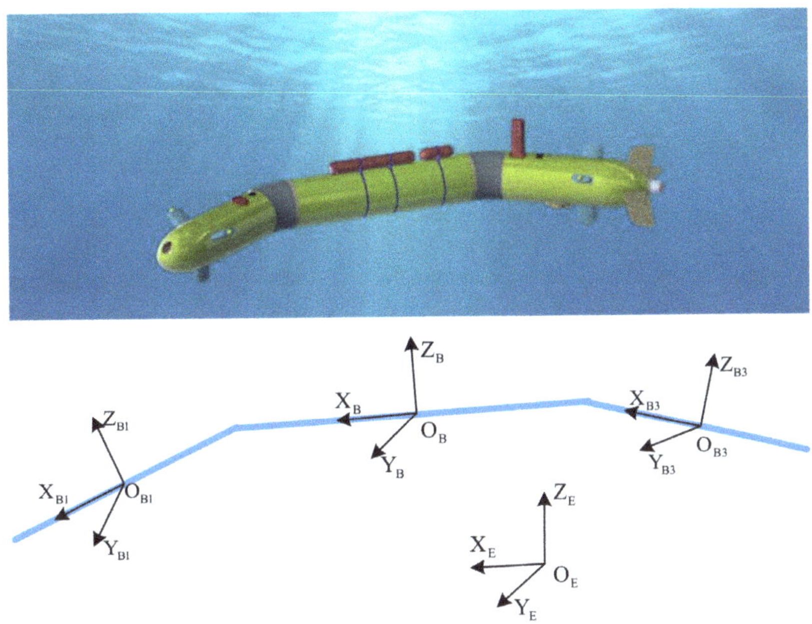

Figure 1. Reference frames of the MJ-AUV.

The technical challenge of the MJ-AUV is the joint design. The rotation of the orthogonal joint requires two motors to cooperate with each other to drive the bevel gears, so as to realize the pitch and yaw motion of the joint. The specific working process is as follows.

(1) As shown in Figure 2, when the two motors rotate in the opposite direction, the gear set is driven to rotate and the pitch motion of the joint can be realized.

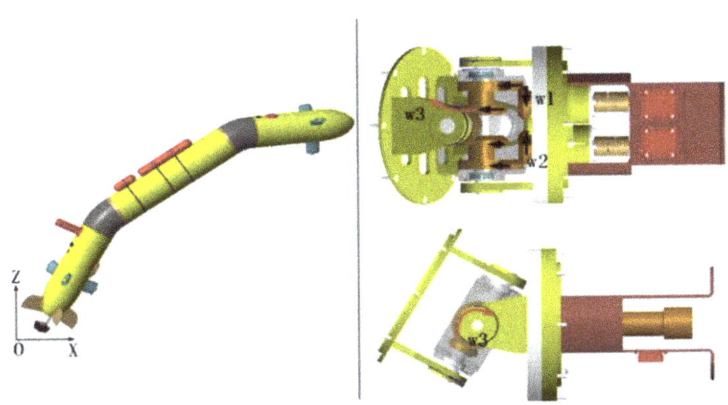

Figure 2. Orthogonal joint pitching motion.

(2) As shown in Figure 3, when the two motors rotate in the same direction, the yaw motion of the joint can be realized.

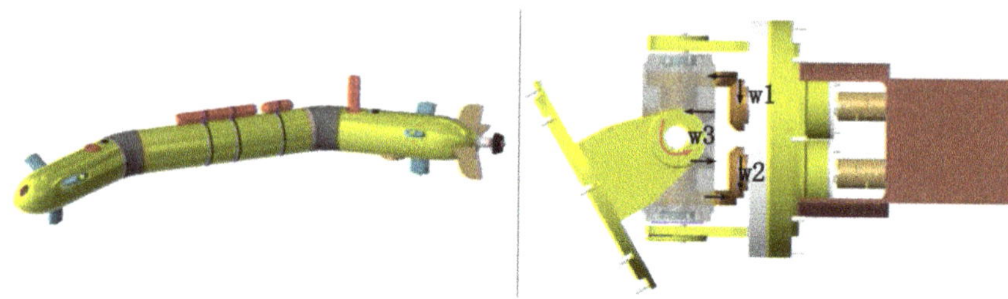

Figure 3. Orthogonal joint yaw motion.

2.2. Assumptions

The MJ-AUV is a complex system with multi-rigid body configuration, high nonlinearity, and strong coupling, which brings great challenges to the modeling work. In order to further reduce the modeling difficulty without losing the generality and accuracy of the model, the following assumptions are proposed:

(1) The vehicle is an ideal multi-rigid body structure, all the forces acting on it can be equivalent to a combined force.
(2) The influence of the rotation of the earth is ignored, that is, the inertial frame is not affected by the force generated by the rotation of the earth.
(3) The attitude adjustment of the orthogonal joint is realized by two motors rotating in the same direction or opposite direction. And the joint angle is a linear mapping relationship with the motor rotation angle, so the joint angle can be directly used as the input of the system.
(4) There is a nonlinear mapping relationship between thrust and speed. In practical engineering, the thrust output of the thruster can be adjusted by the inner loop control. It is assumed that the inner loop control effect is good, and the thrust is defined as the direct input of the system.
(5) The vehicle works in the deep-sea environment, where the movement speed is relatively slow and the joints do not swing frequently. Therefore, it is assumed that the hydrodynamic coefficient of each cabin is only related to its own shape and size.
(6) The center of buoyancy in each cabin coincides with the centroid, and the center of gravity is directly below the centroid.
(7) It is assumed that the density of seawater at different depths does not change much and is approximately constant.

2.3. Coordinate System Definition

Figure 1 defines the inertial frame $O_EX_EY_EZ_E$, body (navigation/control cabin) frame $O_BX_BY_BZ_B$, diversion cabin frame $O_{B1}X_{B1}Y_{B1}Z_{B1}$, and propulsion cabin frame $O_{B3}X_{B3}Y_{B3}Z_{B3}$. The inertial frame is the north-east-and-up coordinate system [30]. O_B is fixed at the centroid position of the navigation/control cabin. O_BX_B runs along the axis of the cabin, O_BZ_B is perpendicular to O_BX_B and upwards, and the establishment of O_BY_B satisfies the right-hand rule. The coordinate system of the diversion cabin and the propulsion cabin is similar to that of the body frame.

2.4. Motion Parameters Definition

(1) Displacement

The displacement of the vehicle mainly includes three parts: surge, sway and heave, which are explained as follows:

Surge X is the projection of the origin position vector of the body frame on X_E, and its direction is the same as X_E.

Sway Y is the projection of the origin position vector of the body frame on Y_E, and its direction is the same as Y_E.

Heave Z is the projection of the origin position vector of the body frame on Z_E, and its direction is the same as Z_E.

(2) Attitude

The attitude angles of the vehicle are determined by the relationship between the body frame and the inertial frame.

Pitch θ is the included angle between X_B and the sea level, when the body is descending, the direction is positive.

Roll φ is the included angle between Z_B and the plumb plane passing through X_B, when the body rolls to the right, the direction is positive.

Yaw ψ is the included angle between the projection of X_B at sea level and X_E, when the body yaws to the left, the direction is positive.

(3) Joint angles

The joint angles are determined by the relationship between the frame of the diversion cabin or propulsion cabin and the body frame. Each orthogonal joint can be used for pitch and yaw control.

The joint pitch angle θ_n is the included angle between X_{B1} or X_{B3} and the plane $X_BO_BY_B$, where $n = 1, 2$ represents the joint n, joint 1 is the joint between the diversion cabin and the navigation/control cabin, and joint 2 is the joint between the navigation/control cabin and the propulsion cabin. When the joint is deflected clockwise, the direction of θ_n is specified as positive.

The joint yaw angle ψ_n is the included angle between the projection of X_{B1} or X_{B3} onto the plane $X_BO_BY_B$ and X_B. When the joint is deflected counterclockwise, the direction of ψ_n is specified as positive.

For modeling convenience, the rotation order of orthogonal joints is defined here. First, it rotates around the Z-axis, then it rotates around the Y-axis of the changed frame.

(4) The linear velocity and angular velocity components of the body coordinate system

u, v, and w are the linear velocities along each axis of the body frame, where u, v, and w are in the same direction as X_B, Y_B, and Z_B, respectively. And p, q, and r represent the component of the attitude angular velocities around each axis of the body frame, which are roll velocity, pitch velocity, and yaw velocity, respectively.

2.5. Kinematic Analysis

The position of the body under the inertial frame is ${}^{E}\boldsymbol{P}_B = \begin{bmatrix} X & Y & Z \end{bmatrix}^T$, and the posture is ${}^{E}\boldsymbol{\Omega}_B = \begin{bmatrix} \varphi & \theta & \psi \end{bmatrix}^T$. The upper left corner of ${}^{E}\boldsymbol{P}_B$ and ${}^{E}\boldsymbol{\Omega}_B$ is the reference frame, which can be omitted when the reference frame is itself, and $\boldsymbol{A}^T$ is the transpose of $\boldsymbol{A}$.

The position of the origin of each cabin ${}^{B}\boldsymbol{P}_i$ is expressed as

$$\begin{cases} \boldsymbol{P}_B = \begin{bmatrix} 0 & 0 & 0 \end{bmatrix}^T \\ {}^{B}\boldsymbol{P}_{B1} = \begin{bmatrix} l_B & 0 & 0 \end{bmatrix}^T + {}^{B}\boldsymbol{R}_{B1}\begin{bmatrix} l_{B1} & 0 & 0 \end{bmatrix}^T \\ {}^{B}\boldsymbol{P}_{B3} = \begin{bmatrix} -l_B & 0 & 0 \end{bmatrix}^T + {}^{B}\boldsymbol{R}_{B3}\begin{bmatrix} -l_{B3} & 0 & 0 \end{bmatrix}^T \end{cases}, \tag{1}$$

where l_i is half the length of each cabin, i = B1, B, B3; and ${}^{B}\boldsymbol{R}_{B1}$ and ${}^{B}\boldsymbol{R}_{B3}$ are the conversion matrices of the diversion cabin frame and the propellant cabin frame to the body frame, respectively, as follows:

$$\begin{aligned} {}^{B}\boldsymbol{R}_{B1} &= \boldsymbol{Rot}(z,\psi_1)\boldsymbol{Rot}(y,\theta_1) \\ {}^{B}\boldsymbol{R}_{B3} &= {}^{B3}\boldsymbol{R}_B{}^{-1} = (\boldsymbol{Rot}(z,\psi_2)\boldsymbol{Rot}(y,\theta_2))^{-1} \end{aligned} \tag{2}$$

$$\begin{array}{l} Rot(z,\psi_n)=\begin{bmatrix} \cos(\psi_n) & -\sin(\psi_n) & 0 \\ \sin(\psi_n) & \cos(\psi_n) & 0 \\ 0 & 0 & 1 \end{bmatrix} \\ Rot(y,\theta_n)=\begin{bmatrix} \cos(\theta_n) & 0 & \sin(\theta_n) \\ 0 & 1 & 0 \\ -\sin(\theta_n) & 0 & \cos(\theta_n) \end{bmatrix} \end{array} \quad (n=1,\ 2), \tag{3}$$

where $\boldsymbol{R}^{-1}$ is the inverse of $\boldsymbol{R}$. Since $\boldsymbol{R}$ is an orthogonal matrix,

$${}^{\mathrm{B}}\boldsymbol{R}_{\mathrm{B3}}{}^{-1} = {}^{\mathrm{B}}\boldsymbol{R}_{\mathrm{B3}}{}^{\mathrm{T}} = Rot(y,\theta_2)^{\mathrm{T}} Rot(z,\psi_2)^{\mathrm{T}}. \tag{4}$$

The angular velocity ${}^{\mathrm{B}}\boldsymbol{\omega}_i$ $(i = B1,\ B,\ \mathrm{B3})$ and linear velocity ${}^{\mathrm{B}}\boldsymbol{v}_i$ $(i = B1,\ B,\ \mathrm{B3})$ are

$$\begin{cases} \boldsymbol{\omega}_{\mathrm{B}} = \begin{bmatrix} p & q & r \end{bmatrix}^{\mathrm{T}} \\ {}^{\mathrm{B}}\boldsymbol{\omega}_{\mathrm{B1}} = \boldsymbol{\omega}_{\mathrm{B}} + \begin{bmatrix} 0 & 0 & \dot{\psi}_1 \end{bmatrix}^{\mathrm{T}} + Rot(z,\psi_1)\begin{bmatrix} 0 & \dot{\theta}_1 & 0 \end{bmatrix}^{\mathrm{T}} \\ {}^{\mathrm{B}}\boldsymbol{\omega}_{\mathrm{B3}} = \boldsymbol{\omega}_{\mathrm{B}} - \begin{bmatrix} 0 & \dot{\theta}_2 & 0 \end{bmatrix}^{\mathrm{T}} - Rot(y,\theta_2)^{\mathrm{T}}\begin{bmatrix} 0 & 0 & \dot{\psi}_2 \end{bmatrix}^{\mathrm{T}} \end{cases}, \tag{5}$$

$$\begin{cases} \boldsymbol{v}_{\mathrm{B}} = \begin{bmatrix} u & v & w \end{bmatrix}^{\mathrm{T}} \\ {}^{\mathrm{B}}\boldsymbol{v}_{\mathrm{B1}} = \boldsymbol{v}_{\mathrm{B}} + \frac{d^{\mathrm{B}}\boldsymbol{P}_{\mathrm{B1}}}{dt} + \boldsymbol{\omega}_{\mathrm{B}} \times {}^{\mathrm{B}}\boldsymbol{P}_{\mathrm{B1}} \\ {}^{\mathrm{B}}\boldsymbol{v}_{\mathrm{B3}} = \boldsymbol{v}_{\mathrm{B}} + \frac{d^{\mathrm{B}}\boldsymbol{P}_{\mathrm{B3}}}{dt} + \boldsymbol{\omega}_{\mathrm{B}} \times {}^{\mathrm{B}}\boldsymbol{P}_{\mathrm{B3}} \end{cases}, \tag{6}$$

where $\dot{\theta}_n$ and $\dot{\psi}_n$ are the pitch and yaw velocities of the joints n (n = 1, 2), respectively. The velocity of the body is expressed in the inertial frame as

$$\begin{cases} {}^{\mathrm{E}}\dot{\boldsymbol{P}}_{\mathrm{B}} = {}^{\mathrm{E}}\boldsymbol{R}_{\mathrm{B}}\boldsymbol{v}_{\mathrm{B}} \\ {}^{\mathrm{E}}\dot{\boldsymbol{\Omega}}_{\mathrm{B}} = {}^{\mathrm{E}}\boldsymbol{G}_{\mathrm{B}}\boldsymbol{\omega}_{\mathrm{B}} \end{cases}, \tag{7}$$

where ${}^{\mathrm{E}}\dot{\boldsymbol{P}}_{\mathrm{B}}$ and ${}^{\mathrm{E}}\dot{\boldsymbol{\Omega}}_{\mathrm{B}}$ are the velocities of the body frame relative to the inertial frame, and ${}^{\mathrm{E}}\boldsymbol{R}_{\mathrm{B}}$ and ${}^{\mathrm{E}}\boldsymbol{G}_{\mathrm{B}}$ are the transformation matrices of the linear velocity and angular velocity from the body frame to the inertial frame, respectively. Tait-Bryan angles (Z-Y-X) rotation transformation is adopted to determine the two matrices as follows.

$$\begin{aligned} {}^{\mathrm{E}}\boldsymbol{R}_{\mathrm{B}} &= Rot(z,\psi)Rot(y,\theta)Rot(x,\varphi) \\ &= \begin{bmatrix} c\theta c\psi & s\varphi s\theta c\psi - c\varphi s\psi & c\varphi s\theta c\psi + s\varphi s\psi \\ c\theta s\psi & s\varphi s\theta s\psi + c\varphi c\psi & c\varphi s\theta s\psi - s\varphi c\psi \\ -s\theta & s\varphi c\theta & c\varphi c\theta \end{bmatrix}, \end{aligned} \tag{8}$$

$${}^{\mathrm{E}}\mathbf{G}_{\mathrm{B}} = \begin{bmatrix} 1 & s\varphi t\theta & c\varphi t\theta \\ 0 & c\varphi & -s\varphi \\ 0 & s\varphi/c\theta & c\varphi/c\theta \end{bmatrix} \tag{9}$$

for any $j = \varphi,\ \theta,\ \psi,\ sj = \sin j,\ cj = \cos j,\ tj = \tan j$; $Rot(z,\psi)$ and $Rot(y,\theta)$ are similar to Equation (3), and

$$Rot(x,\varphi) = \begin{bmatrix} 1 & 0 & 0 \\ 0 & \cos(\phi) & -\sin(\phi) \\ 0 & \sin(\phi) & \cos(\phi) \end{bmatrix}, \tag{10}$$

Equation (9) can be obtained by solving

$$\begin{bmatrix} p \\ q \\ r \end{bmatrix} = \begin{bmatrix} \dot{\phi} \\ 0 \\ 0 \end{bmatrix} + Rot(x,\phi)^T \begin{bmatrix} 0 \\ \dot{\theta} \\ 0 \end{bmatrix} + Rot(x,\phi)^T Rot(y,\theta)^T \begin{bmatrix} 0 \\ 0 \\ \dot{\psi} \end{bmatrix}. \tag{11}$$

Subject to the mechanical limit, the pitching angle will not reach $\pm 90°$, so ${}^{\mathrm{E}}G_{\mathrm{B}}$ will not be in a singular state.

The angular acceleration ${}^{\mathrm{B}}\boldsymbol{\alpha}_i$ $(i = B1,\ \mathrm{B},\ \mathrm{B3})$ and linear acceleration ${}^{\mathrm{B}}\boldsymbol{a}_i$ $(i = B1,\ \mathrm{B},\ \mathrm{B3})$ are expressed as follows:

$$\begin{cases} \boldsymbol{\alpha}_{\mathrm{B}} = \begin{bmatrix} \dot{p} & \dot{q} & \dot{r} \end{bmatrix}^{\mathrm{T}} \\ {}^{\mathrm{B}}\boldsymbol{\alpha}_{\mathrm{B1}} = \frac{d^{\mathrm{B}}\boldsymbol{\omega}_{\mathrm{B1}}}{dt} + \boldsymbol{\omega}_{\mathrm{B}} \times {}^{\mathrm{B}}\boldsymbol{\omega}_{\mathrm{B1}} \\ {}^{\mathrm{B}}\boldsymbol{\alpha}_{\mathrm{B3}} = \frac{d^{\mathrm{B}}\boldsymbol{\omega}_{\mathrm{B3}}}{dt} + \boldsymbol{\omega}_{\mathrm{B}} \times {}^{\mathrm{B}}\boldsymbol{\omega}_{\mathrm{B3}} \end{cases}, \tag{12}$$

$$\begin{cases} \boldsymbol{a}_{\mathrm{B}} = \frac{d\boldsymbol{v}_{\mathrm{B}}}{dt} + \boldsymbol{\omega}_{\mathrm{B}} \times \boldsymbol{v}_{\mathrm{B}} \\ {}^{\mathrm{B}}\boldsymbol{a}_{\mathrm{B1}} = \frac{d^{\mathrm{B}}\boldsymbol{v}_{\mathrm{B1}}}{dt} + \boldsymbol{\omega}_{\mathrm{B}} \times {}^{\mathrm{B}}\boldsymbol{v}_{\mathrm{B1}} \\ {}^{\mathrm{B}}\boldsymbol{a}_{\mathrm{B3}} = \frac{d^{\mathrm{B}}\boldsymbol{v}_{\mathrm{B3}}}{dt} + \boldsymbol{\omega}_{\mathrm{B}} \times {}^{\mathrm{B}}\boldsymbol{v}_{\mathrm{B3}} \end{cases}, \tag{13}$$

where $\dot{p}$, $\dot{q}$, and $\dot{r}$ are the angular accelerations of the roll, pitch, and yaw of the body, respectively.

2.6. Dynamic Analysis

In the process of moving, the MJ-AUV is mainly subject to the fluid drag, the inertial hydrodynamic force caused by additional mass, buoyancy, gravity, thrust, and interaction forces between joints.

2.6.1. Hydrodynamic Analysis

Figure 1 shows the structure of the MJ-AUV equipped with multiple sensors. Notably, it is not a regular cylinder. Considering the influence of the pressure drag and the trailing vortex shedding effect, the drag suffered by each cabin is expressed as follows:

$${}^{\mathrm{B}}\boldsymbol{F}_{\mathrm{d}i} = {}^{\mathrm{B}}\boldsymbol{R}_i\left(-\boldsymbol{C}_{\mathrm{D}i}^2 \mathrm{sgn}(\boldsymbol{v}_i)\boldsymbol{v}_i{}^2 - \boldsymbol{C}_{\mathrm{D}i}^1 \boldsymbol{v}_i\right),\ (i = \mathrm{B1},\ \mathrm{B},\ \mathrm{B3}) \tag{14}$$

with

$$\boldsymbol{C}_{\mathrm{D}i}^2 = \begin{bmatrix} c_{\mathrm{D}ix}^2 & 0 & 0 \\ 0 & c_{\mathrm{D}iy}^2 & 0 \\ 0 & 0 & c_{\mathrm{D}iz}^2 \end{bmatrix},\ (i = \mathrm{B1},\ \mathrm{B},\ \mathrm{B3})$$

$$\boldsymbol{C}_{\mathrm{D}i}^1 = \begin{bmatrix} c_{\mathrm{D}ix}^1 & 0 & 0 \\ 0 & c_{\mathrm{D}iy}^1 & 0 \\ 0 & 0 & c_{\mathrm{D}iz}^1 \end{bmatrix},\ (i = \mathrm{B1},\ \mathrm{B},\ \mathrm{B3})$$

$$\mathrm{sgn}(s) = \begin{cases} 1, & s > 0 \\ 0, & s = 0 \\ -1, & s < 0 \end{cases},$$

$$\boldsymbol{v}_i = {}^{\mathrm{B}}\boldsymbol{R}_i{}^{\mathrm{T}\mathrm{B}}\boldsymbol{v}_i,\ (i = \mathrm{B1},\ \mathrm{B},\ \mathrm{B3})$$

where $c_{\mathrm{D}ix}^2$, $c_{\mathrm{D}iy}^2$, and $c_{\mathrm{D}iz}^2$, and $c_{\mathrm{D}ix}^1$, $c_{\mathrm{D}iy}^1$, and $c_{\mathrm{D}iz}^1$ are the quadratic and the first-order coefficients of drag on 3D linear velocity, respectively.

When the MJ-AUV travels with variable speed motion, it forms a relative acceleration motion with the surrounding water bodies, causing the additional mass effect and producing the effect opposite to the direction of acceleration, which can be expressed by

$${}^{\mathrm{B}}\boldsymbol{F}_{\mathrm{a}i} = -{}^{\mathrm{B}}\boldsymbol{R}_i(\boldsymbol{\lambda}_{\mathrm{m}i}\boldsymbol{a}_i),\ \boldsymbol{M}_{\mathrm{a}i} = -\boldsymbol{\lambda}_{\mathrm{J}i}\boldsymbol{\alpha}_i,\ (i = \mathrm{B1},\ \mathrm{B},\ \mathrm{B3}), \tag{15}$$

with

$$\boldsymbol{\lambda}_i = \begin{bmatrix} \boldsymbol{\lambda}_{\mathrm{m}i} & 0_{3\times3} \\ 0_{3\times3} & \boldsymbol{\lambda}_{\mathrm{J}i} \end{bmatrix} = \begin{bmatrix} \lambda_{i11} & 0 & 0 & 0 & 0 & 0 \\ 0 & \lambda_{i22} & 0 & 0 & 0 & 0 \\ 0 & 0 & \lambda_{i33} & 0 & 0 & 0 \\ 0 & 0 & 0 & \lambda_{i44} & 0 & 0 \\ 0 & 0 & 0 & 0 & \lambda_{i55} & 0 \\ 0 & 0 & 0 & 0 & 0 & \lambda_{i66} \end{bmatrix}, \ (i = \mathrm{B1,\ B,\ B3})$$

$$\boldsymbol{a}_i = {}^{\mathrm{B}}\boldsymbol{R}_i{}^{\mathrm{TB}}\boldsymbol{a}_i, \ (i = \mathrm{B1,\ B,\ B3})$$

and $\boldsymbol{\lambda}_i$ is the added mass matrix of each cabin.

The buoyancy of each cabin of the MJ-AUV is equal to the gravity of the water discharged from the cabin. Here, assuming that the density of seawater is almost constant at different depths. Thus, when the MJ-AUV is fully immersed in seawater, the buoyancy of each cabin is

$$\begin{aligned} {}^{\mathrm{B}}\boldsymbol{F}_{\mathrm{b}i} &= {}^{\mathrm{E}}\boldsymbol{R}_{\mathrm{B}}{}^{\mathrm{T}}\begin{bmatrix} 0 & 0 & {}^{\mathrm{E}}F_{\mathrm{b}i} \end{bmatrix}^{\mathrm{T}}, \ (i = \mathrm{B1,\ B,\ B3}) \\ {}^{\mathrm{E}}F_{\mathrm{b}i} &= \rho_{\mathrm{liquid}} V_i g, \ (i = \mathrm{B1,\ B,\ B3}) \end{aligned} \tag{16}$$

where ρ_{liquid} is the density of seawater, V_i is the volume of each cabin, and g is the gravitational acceleration. The buoyancy of each cabin acts at the center, so the moment of which is 0.

2.6.2. Gravity and Gravitational Moment

In the air, the gravity on each cabin is expressed in the body frame as:

$$\begin{aligned} {}^{\mathrm{B}}\boldsymbol{G}_i &= {}^{\mathrm{E}}\boldsymbol{R}_{\mathrm{B}}{}^{\mathrm{TE}}\boldsymbol{G}_i, \ (i = \mathrm{B1,\ B,\ B3}) \\ {}^{\mathrm{E}}\boldsymbol{G}_i &= \begin{bmatrix} 0 & 0 & m_i g \end{bmatrix}^{\mathrm{T}}, \ (i = \mathrm{B1,\ B,\ B3}) \end{aligned} \tag{17}$$

where m_i is the mass of each cabin.

The center of gravity is directly below the centroid, so each cabin is affected by the gravitational moment as:

$$\boldsymbol{M}_{\mathrm{g}i} = \begin{bmatrix} 0 & 0 & -l_{\mathrm{c}} \end{bmatrix}^{\mathrm{T}} \times \left({}^{\mathrm{B}}\boldsymbol{R}_i{}^{\mathrm{TB}}\boldsymbol{G}_i\right), \ (i = \mathrm{B1,\ B,\ B3}) \tag{18}$$

where l_{c} represents the distance between the center of gravity and the centroid of each cabin.

2.6.3. Thrust Analysis

The propeller is installed at the rear of the propulsion cabin of the MJ-AUV and is a one-way force. The direction is forward along the axial direction of the propulsion cabin. Therefore, the thrust does not produce a torque effect on the propulsion cabin. The thrust is expressed in the body frame as:

$${}^{\mathrm{B}}\boldsymbol{F}_{\mathrm{th}} = {}^{\mathrm{B}}\boldsymbol{R}_{\mathrm{B3}}\boldsymbol{F}_{\mathrm{th}}, \tag{19}$$

with $\boldsymbol{F}_{\mathrm{th}} = \begin{bmatrix} F_{\mathrm{t}} & 0 & 0 \end{bmatrix}^{\mathrm{T}}$ and F_{t} is the thrust from the propeller.

2.6.4. Dynamical Equation

According to Newton's second law, the force analysis is shown in the following formula.

$$\begin{cases} {}^{\mathrm{B}}\boldsymbol{F}_{\mathrm{bB1}} + {}^{\mathrm{B}}\boldsymbol{F}_{\mathrm{dB1}} + {}^{\mathrm{B}}\boldsymbol{F}_{\mathrm{aB1}} - {}^{\mathrm{B}}\boldsymbol{G}_{\mathrm{B1}} + {}^{\mathrm{B}}\boldsymbol{F}_{\mathrm{B}\to\mathrm{B1}} = m_{\mathrm{B1}}\boldsymbol{I}^{\mathrm{B}}\boldsymbol{a}_{\mathrm{B1}} \\ \boldsymbol{F}_{\mathrm{bB}} + \boldsymbol{F}_{\mathrm{dB}} + \boldsymbol{F}_{\mathrm{aB}} - \boldsymbol{G}_{\mathrm{B}} - {}^{\mathrm{B}}\boldsymbol{F}_{\mathrm{B}\to\mathrm{B1}} + {}^{\mathrm{B}}\boldsymbol{F}_{\mathrm{B3}\to\mathrm{B}} = m_{\mathrm{B}}\boldsymbol{I}\boldsymbol{a}_{\mathrm{B}} \\ {}^{\mathrm{B}}\boldsymbol{F}_{\mathrm{bB3}} + {}^{\mathrm{B}}\boldsymbol{F}_{\mathrm{dB3}} + {}^{\mathrm{B}}\boldsymbol{F}_{\mathrm{aB3}} - {}^{\mathrm{B}}\boldsymbol{G}_{\mathrm{B3}} - {}^{\mathrm{B}}\boldsymbol{F}_{\mathrm{B3}\to\mathrm{B}} + {}^{\mathrm{B}}\boldsymbol{F}_{\mathrm{th}} = m_{\mathrm{B3}}\boldsymbol{I}^{\mathrm{B}}\boldsymbol{a}_{\mathrm{B3}} \end{cases}, \quad (20)$$

where ${}^{\mathrm{B}}\boldsymbol{F}_{\mathrm{B}\to\mathrm{B1}}$ represents the force of the navigation/control cabin on the diversion cabin in the body frame, ${}^{\mathrm{B}}\boldsymbol{F}_{\mathrm{B3}\to\mathrm{B}}$ stands for the force of the propulsion cabin on the navigation/control cabin in the body frame, and $\boldsymbol{I}$ is a 3×3 identity matrix. By adding the three expressions in Equation (20), the interaction force between joints can be eliminated, i.e.

$$\begin{aligned} &{}^{\mathrm{B}}\boldsymbol{F}_{\mathrm{bB1}} + {}^{\mathrm{B}}\boldsymbol{F}_{\mathrm{dB1}} + {}^{\mathrm{B}}\boldsymbol{F}_{\mathrm{aB1}} - {}^{\mathrm{B}}\boldsymbol{G}_{\mathrm{B1}} + \boldsymbol{F}_{\mathrm{bB}} + \boldsymbol{F}_{\mathrm{dB}} + \boldsymbol{F}_{\mathrm{aB}} - \boldsymbol{G}_{\mathrm{B}} + {}^{\mathrm{B}}\boldsymbol{F}_{\mathrm{bB3}} + \\ &{}^{\mathrm{B}}\boldsymbol{F}_{\mathrm{dB3}} + {}^{\mathrm{B}}\boldsymbol{F}_{\mathrm{aB3}} - {}^{\mathrm{B}}\boldsymbol{G}_{\mathrm{B3}} + {}^{\mathrm{B}}\boldsymbol{F}_{\mathrm{th}} - m_{\mathrm{B1}}\boldsymbol{I}^{\mathrm{B}}\boldsymbol{a}_{\mathrm{B1}} - m_{\mathrm{B}}\boldsymbol{I}\boldsymbol{a}_{\mathrm{B}} - m_{\mathrm{B3}}\boldsymbol{I}^{\mathrm{B}}\boldsymbol{a}_{\mathrm{B3}} = 0 \end{aligned} \quad (21)$$

According to the torque balance principle, the torque analysis is shown in the following formula.

$$\begin{cases} \boldsymbol{M}_{\mathrm{aB1}} + {}^{\mathrm{B1}}\boldsymbol{M}_1 - \boldsymbol{M}_{\mathrm{gB1}} + {}^{\mathrm{B1}}\boldsymbol{M}_{\mathrm{B}\to\mathrm{B1}} = \boldsymbol{J}_{\mathrm{B1}}\boldsymbol{\alpha}_{\mathrm{B1}} \\ \boldsymbol{M}_{\mathrm{aB}} + {}^{\mathrm{B}}\boldsymbol{M}_2 - {}^{\mathrm{B}}\boldsymbol{M}_1 - \boldsymbol{M}_{\mathrm{gB}} - {}^{\mathrm{B}}\boldsymbol{M}_{\mathrm{B}\to\mathrm{B1}} + {}^{\mathrm{B}}\boldsymbol{M}_{\mathrm{B3}\to\mathrm{B}} = \boldsymbol{J}_{\mathrm{B}}\boldsymbol{\alpha}_{\mathrm{B}} \\ \boldsymbol{M}_{\mathrm{aB3}} - {}^{\mathrm{B3}}\boldsymbol{M}_2 - \boldsymbol{M}_{\mathrm{gB3}} - {}^{\mathrm{B3}}\boldsymbol{M}_{\mathrm{B3}\to\mathrm{B}} = \boldsymbol{J}_{\mathrm{B3}}\boldsymbol{\alpha}_{\mathrm{B3}} \end{cases}, \quad (22)$$

with

$${}^{\mathrm{B1}}\boldsymbol{M}_{\mathrm{B}\to\mathrm{B1}} = \begin{bmatrix} -l_{\mathrm{B1}} & 0 & 0 \end{bmatrix}^{\mathrm{T}} \times \left({}^{\mathrm{B}}\boldsymbol{R}_{\mathrm{B1}}{}^{\mathrm{TB}}\boldsymbol{F}_{\mathrm{B}\to\mathrm{B1}}\right),$$

$${}^{\mathrm{B}}\boldsymbol{M}_{\mathrm{B}\to\mathrm{B1}} = \begin{bmatrix} l_{\mathrm{B1}} & 0 & 0 \end{bmatrix}^{\mathrm{T}} \times {}^{\mathrm{B}}\boldsymbol{F}_{\mathrm{B}\to\mathrm{B1}},$$

$${}^{\mathrm{B}}\boldsymbol{M}_{\mathrm{B3}\to\mathrm{B}} = \begin{bmatrix} -l_{\mathrm{B}} & 0 & 0 \end{bmatrix}^{\mathrm{T}} \times {}^{\mathrm{B}}\boldsymbol{F}_{\mathrm{B3}\to\mathrm{B}},$$

$${}^{\mathrm{B3}}\boldsymbol{M}_{\mathrm{B3}\to\mathrm{B}} = \begin{bmatrix} l_{\mathrm{B3}} & 0 & 0 \end{bmatrix}^{\mathrm{T}} \times \left({}^{\mathrm{B}}\boldsymbol{R}_{\mathrm{B3}}{}^{\mathrm{TB}}\boldsymbol{F}_{\mathrm{B3}\to\mathrm{B}}\right),$$

$$\boldsymbol{J}_i = \begin{bmatrix} J_{ixx} & 0 & 0 \\ 0 & J_{iyy} & 0 \\ 0 & 0 & J_{izz} \end{bmatrix}, \ (i = B1,\ \mathrm{B},\ \mathrm{B3})$$

$$\boldsymbol{\alpha}_i = {}^{\mathrm{B}}\boldsymbol{R}_i{}^{\mathrm{TB}}\boldsymbol{\alpha}_i, \ (i = B1,\ \mathrm{B},\ \mathrm{B3})$$

where ${}^{\mathrm{B1}}\boldsymbol{M}_{\mathrm{B}\to\mathrm{B1}}$ is the torque generated by the navigation/control cabin on the diversion cabin expressed in the diversion cabin frame; the definitions of ${}^{\mathrm{B}}\boldsymbol{M}_{\mathrm{B}\to\mathrm{B1}}$, ${}^{\mathrm{B}}\boldsymbol{M}_{\mathrm{B3}\to\mathrm{B}}$, and ${}^{\mathrm{B3}}\boldsymbol{M}_{\mathrm{B3}\to\mathrm{B}}$ are similar to ${}^{\mathrm{B1}}\boldsymbol{M}_{\mathrm{B}\to\mathrm{B1}}$; $\boldsymbol{J}_i$ is the matrix of the moment of inertia of each cabin; and ${}^{\mathrm{B}}\boldsymbol{F}_{\mathrm{B}\to\mathrm{B1}}$ and ${}^{\mathrm{B}}\boldsymbol{F}_{\mathrm{B3}\to\mathrm{B}}$ can be obtained from Equation (20).

All the variables in Equation (22) are expressed in the body frame,

$$\begin{cases} {}^{\mathrm{B}}\boldsymbol{R}_{\mathrm{B1}}\left(\boldsymbol{M}_{\mathrm{aB1}} - \boldsymbol{M}_{\mathrm{gB1}} + {}^{\mathrm{B1}}\boldsymbol{M}_{\mathrm{B}\to\mathrm{B1}}\right) + {}^{\mathrm{B}}\boldsymbol{M}_1 = {}^{\mathrm{B}}\boldsymbol{R}_{\mathrm{B1}}\boldsymbol{J}_{\mathrm{B1}}\boldsymbol{\alpha}_{\mathrm{B1}} \\ \boldsymbol{M}_{\mathrm{aB}} + {}^{\mathrm{B}}\boldsymbol{M}_2 - {}^{\mathrm{B}}\boldsymbol{M}_1 - \boldsymbol{M}_{\mathrm{gB}} - {}^{\mathrm{B}}\boldsymbol{M}_{\mathrm{B}\to\mathrm{B1}} + {}^{\mathrm{B}}\boldsymbol{M}_{\mathrm{B3}\to\mathrm{B}} = \boldsymbol{J}_{\mathrm{B}}\boldsymbol{\alpha}_{\mathrm{B}} \\ {}^{\mathrm{B}}\boldsymbol{R}_{\mathrm{B3}}\left(\boldsymbol{M}_{\mathrm{aB3}} - \boldsymbol{M}_{\mathrm{gB3}} - {}^{\mathrm{B3}}\boldsymbol{M}_{\mathrm{B3}\to\mathrm{B}}\right) - {}^{\mathrm{B}}\boldsymbol{M}_2 = {}^{\mathrm{B}}\boldsymbol{R}_{\mathrm{B3}}\boldsymbol{J}_{\mathrm{B3}}\boldsymbol{\alpha}_{\mathrm{B3}} \end{cases}. \quad (23)$$

The attitude of the MJ-AUV is controlled by changing the joint angle. By adding the three equations in (23), the following equation can be obtained:

$$\begin{aligned} &{}^{\mathrm{B}}\boldsymbol{R}_{\mathrm{B1}}\left(\boldsymbol{M}_{\mathrm{aB1}} - \boldsymbol{M}_{\mathrm{gB1}} + {}^{\mathrm{B1}}\boldsymbol{M}_{\mathrm{B}\to\mathrm{B1}} - \boldsymbol{J}_{\mathrm{B1}}\boldsymbol{\alpha}_{\mathrm{B1}}\right) + \boldsymbol{M}_{\mathrm{aB}} - \boldsymbol{M}_{\mathrm{gB}} - {}^{\mathrm{B}}\boldsymbol{M}_{\mathrm{B}\to\mathrm{B1}} + \\ &{}^{\mathrm{B}}\boldsymbol{M}_{\mathrm{B3}\to\mathrm{B}} - \boldsymbol{J}_{\mathrm{B}}\boldsymbol{\alpha}_{\mathrm{B}} + {}^{\mathrm{B}}\boldsymbol{R}_{\mathrm{B3}}\left(\boldsymbol{M}_{\mathrm{aB3}} - \boldsymbol{M}_{\mathrm{gB3}} - {}^{\mathrm{B3}}\boldsymbol{M}_{\mathrm{B3}\to\mathrm{B}} - \boldsymbol{J}_{\mathrm{B3}}\boldsymbol{\alpha}_{\mathrm{B3}}\right) = 0 \end{aligned} \quad (24)$$

The dynamic models of the MJ-AUV, i.e., Equation (7), Equation (21), and Equation (24), can be expressed as the state space equation as follows:

$$\mathcal{M}\dot{X} = \eta(X, U),$$
$$\mathcal{M} = \begin{bmatrix} I_{6\times 6} & 0_{6\times 6} \\ 0_{6\times 6} & \mathcal{M}_{22} \end{bmatrix},$$
$$\mathcal{M}_{22} = \begin{bmatrix} \frac{\partial F_{\text{total}}}{\partial \dot{u}} & \frac{\partial F_{\text{total}}}{\partial \dot{v}} & \frac{\partial F_{\text{total}}}{\partial \dot{w}} & \frac{\partial F_{\text{total}}}{\partial \dot{p}} & \frac{\partial F_{\text{total}}}{\partial \dot{q}} & \frac{\partial F_{\text{total}}}{\partial \dot{r}} \\ \frac{\partial M_{\text{total}}}{\partial \dot{u}} & \frac{\partial M_{\text{total}}}{\partial \dot{v}} & \frac{\partial M_{\text{total}}}{\partial \dot{w}} & \frac{\partial M_{\text{total}}}{\partial \dot{p}} & \frac{\partial M_{\text{total}}}{\partial \dot{q}} & \frac{\partial M_{\text{total}}}{\partial \dot{r}} \end{bmatrix}, \tag{25}$$

where $\mathcal{M}$ is a 12×12 inertial matrix, $X = \begin{bmatrix} X & Y & Z & \varphi & \theta & \psi & u & v & w & p & q & r \end{bmatrix}^{\mathrm{T}}$, $U = \begin{bmatrix} \theta_1 & \theta_2 & \psi_1 & \psi_2 & F_t \end{bmatrix}^{\mathrm{T}}$, and $\eta(X, U)$ represents the function about X and U.

Equation (25) can be rewritten as

$$\dot{X} = \mathcal{M}^{-1}\eta(X, U). \tag{26}$$

The dynamic model is an analytical model that can be used to study the motion characteristics of the MJ-AUV and design a modern controller based on this model.

3. Attitude Controller Design

By adjusting the pitch and yaw angles of joints 1 and 2, the relative attitude of each cabin is adjusted, the hydrodynamic appearance of the MJ-AUV is changed, and then the attitude angle of the body is adjusted. Because of the mechanical structure characteristics, joints can only change with one degree of freedom at the same time. To realize pitch and yaw control of the body at the same time, stipulating that joint 1 is used for the yaw attitude adjustment, and joint 2 is used for the pitch attitude adjustment, namely, $U = \begin{bmatrix} 0 & \theta_2 & \psi_1 & 0 \end{bmatrix}^{\mathrm{T}}$. The MJ-AUV is a complex system with high nonlinear and strong coupling. To reduce the coupling degree and operation cost of the control system and improve the anti-disturbance performance, internal forces, coupling factors, and external disturbance are regarded as total disturbance and establish the subsystem model of the pitch and yaw as follows:

$$\begin{cases} \ddot{\theta} = f_\theta + b_{11}\psi_1 + b_{12}\theta_2 \\ \ddot{\psi} = f_\psi + b_{21}\psi_1 + b_{22}\theta_2 \end{cases}, \tag{27}$$

where $\ddot{\theta}$ is the pitch acceleration of the body, $\ddot{\psi}$ is the yaw acceleration, $b_{ij}(i, j = 1, 2)$ is the control coefficient of joints 1 and 2 in pitch and yaw motion equations, and f_θ and f_ψ are nonlinear total perturbation functions associated with X and the input coupling terms.

The pitch and yaw attitude control of the MJ-AUV mainly includes the total disturbance observation compensation and LQR control of each subsystem. The designed controller structure is shown in Figure 4.

3.1. Linear Extended State Observer Design

The LESO is used to estimate the states and disturbances of the system that cannot be measured directly. Taking the pitch attitude subsystem as an example, the process of LESO design is as follows.

Let $x_1 = \theta$, $x_2 = \dot{\theta}$, $x_3 = f_\theta$, $u_1 = \psi_1$, and $u_2 = \theta_2$; then, the pitch attitude subsystem can be rewritten as:

$$\begin{aligned} \dot{x} &= Ax + Bu + Eh \\ y &= Cx + Du \end{aligned}, \tag{28}$$

where $A = \begin{bmatrix} 0 & 1 & 0 \\ 0 & 0 & 1 \\ 0 & 0 & 0 \end{bmatrix}, B = \begin{bmatrix} 0 & 0 \\ b_{11} & b_{12} \\ 0 & 0 \end{bmatrix}, C = \begin{bmatrix} 1 & 0 & 0 \end{bmatrix}, D = \begin{bmatrix} 0 & 0 \end{bmatrix}, E = \begin{bmatrix} 0 \\ 0 \\ 1 \end{bmatrix}$, $x = \begin{bmatrix} x_1 & x_2 & x_3 \end{bmatrix}^{\mathrm{T}}, u = \begin{bmatrix} u_1 & u_2 \end{bmatrix}^{\mathrm{T}}$, and $h = \dot{f}_\theta$ is unknown but bounded.

Define $\hat{x} = \begin{bmatrix} \hat{x}_1 & \hat{x}_2 & \hat{x}_3 \end{bmatrix}^{\mathrm{T}}$ and $\hat{y}$ as the observed values of x and y. Consequently, the LESO can be constructed as

$$\begin{aligned} \dot{\hat{x}} &= A\hat{x} + Bu + L(y - \hat{y}) \\ \hat{y} &= C\hat{x} + Du \end{aligned}, \tag{29}$$

where $L = \begin{bmatrix} \beta_1 & \beta_2 & \beta_3 \end{bmatrix}^{\mathrm{T}}$ is the observer gain vector.

Define $\tilde{x} = x - \hat{x}$ is the state estimation error. By combining (28) and (29), can be obtained

$$\begin{aligned} \dot{\hat{x}} &= A\hat{x} + Bu + L(y - \hat{y}) \\ \hat{y} &= C\hat{x} + Du \\ \tilde{A} &= A - LC = \begin{bmatrix} -\beta_1 & 1 & 0 \\ -\beta_2 & 0 & 1 \\ -\beta_3 & 0 & 0 \end{bmatrix}. \end{aligned}, \tag{30}$$

According to [29], through the pole assignment method, all poles are placed at $-\omega_0$, then $\beta_1 = 3\omega_0$, $\beta_2 = 3{\omega_0}^2$, and $\beta_3 = {\omega_0}^3$. $\omega_0(\omega_0 > 0)$ is the bandwidth of the state observer. The estimated state of the observer can be adjusted by adjusting ω_0, especially the estimated value $\hat{f}_\theta = \hat{x}_3$ of the total disturbance f_θ. The design of the LESO for the yaw subsystem is similar to that for the pitch subsystem.

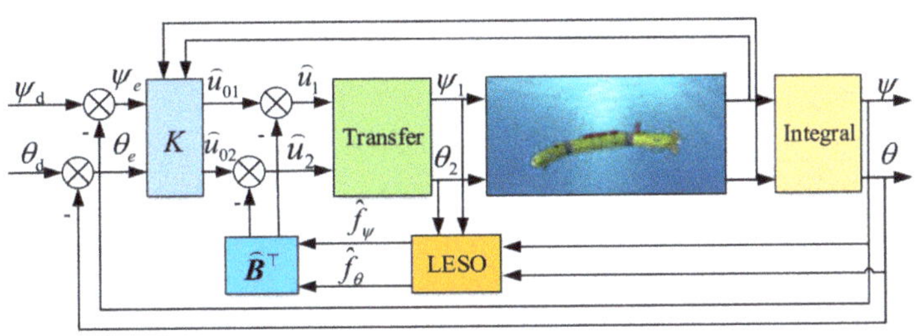

Figure 4. The structure of the LQR decoupling control method based on LESO.

3.2. Control Law Design

To reduce the input coupling degree of the system, the virtual control quantity $\widehat{u} = \begin{bmatrix} b_{11} & b_{12} \\ b_{21} & b_{22} \end{bmatrix} u$ is introduced.

θ_d and ψ_d are defined as the designed pitch and yaw angles, respectively. Then, the feedback errors are $e_\theta = \theta_d - \theta$, $e_\psi = \psi_d - \psi$. Let $z_1 = e_\theta$, $z_2 = \dot{\theta}$, $z_3 = e_\psi$, and $z_4 = \dot{\psi}$; then, the state equation of the pitch and yaw attitude control of the MJ-AUV is

$$\dot{z} = \widehat{A}z + \widehat{B}\widehat{u} + \hat{f}, \tag{31}$$

where

$$\widehat{A} = \begin{bmatrix} 0 & -1 & 0 & 0 \\ 0 & 0 & 0 & 0 \\ 0 & 0 & 0 & -1 \\ 0 & 0 & 0 & 0 \end{bmatrix}, \widehat{B} = \begin{bmatrix} 0 & 0 \\ 0 & 1 \\ 0 & 0 \\ 1 & 0 \end{bmatrix},$$

$\mathbf{z} = \begin{bmatrix} z_1 & z_2 & z_3 & z_4 \end{bmatrix}^{\mathrm{T}}$, $\hat{\mathbf{f}} = \begin{bmatrix} 0 & \hat{f}_\theta & 0 & \hat{f}_\psi \end{bmatrix}^{\mathrm{T}}$, $\hat{f}_\theta$ and $\hat{f}_\psi$ respectively represent the total disturbance of the pitch and yaw subsystems estimated by LESO.

The control law mainly includes disturbance compensation $\widehat{\mathbf{u}}_{\mathrm{f}}$ and linear state feedback $\widehat{\mathbf{u}}_0$, and $\widehat{\mathbf{u}} = \widehat{\mathbf{u}}_{\mathrm{f}} + \widehat{\mathbf{u}}_0$.

The disturbance compensation control law is

$$\widehat{\mathbf{u}}_{\mathrm{f}} = -\widehat{\mathbf{B}}^{\mathrm{T}}\hat{\mathbf{f}}. \tag{32}$$

Substituting Equation (32) into Equation (31) obtains the following:

$$\dot{\mathbf{z}} = \widehat{\mathbf{A}}\mathbf{z} + \widehat{\mathbf{B}}\widehat{\mathbf{u}}_0. \tag{33}$$

Now, the problem is to design the controller $\widehat{\mathbf{u}}_0$, and the linear state feedback control law can be designed as

$$\widehat{\mathbf{u}}_0 = -\mathbf{K}\mathbf{z}, \tag{34}$$

with $\mathbf{K} = \begin{bmatrix} K_{11} & K_{12} & K_{13} & K_{14} \\ K_{21} & K_{22} & K_{23} & K_{24} \end{bmatrix}$ as the linear feedback gain matrix.

The feedback gain matrix has eight parameters. In this paper, LQR can be optimized by LQR to reduce the difficulty of parameter adjustment and obtain the optimal control law suitable for the control target by minimizing the performance index. The LQR performance index function is selected as

$$J = \int_0^\infty \left(\mathbf{z}^{\mathrm{T}}\mathbf{Q}\mathbf{z} + \widehat{\mathbf{u}}_0^{\mathrm{T}}\mathbf{R}\widehat{\mathbf{u}}_0 \right) dt, \tag{35}$$

where $\mathbf{Q}$ is the positive semidefinite matrix and $\mathbf{R}$ is the positive definite matrix. The former is the penalty function of the system state error, and the latter is the penalty function of the system input state. The Riccati equation corresponding to the performance index function is

$$\widehat{\mathbf{A}}^{\mathrm{T}}\mathbf{P} + \mathbf{P}\widehat{\mathbf{A}} + \mathbf{Q} - \mathbf{P}\widehat{\mathbf{B}}\mathbf{R}^{-1}\widehat{\mathbf{B}}^{\mathrm{T}}\mathbf{P} = 0, \tag{36}$$

where $\mathbf{P}$ is the solution of the Riccati equation.

In the case of the Riccati equation, the linear feedback gain matrix is

$$\mathbf{K} = \mathbf{R}^{-1}\mathbf{B}^{\mathrm{T}}\mathbf{P}. \tag{37}$$

In this case, $\mathbf{K}$ is related to $\mathbf{Q}$ and $\mathbf{R}$.

3.3. Stability Analysis

In this part, the stability of the LESO and the closed-loop system are studied according to the idea of reference [31].

A state space equation is defined in the form shown in Equation (38).

$$\dot{\mathbf{c}} = \mathbf{H}\mathbf{c} + \mathbf{o}, \tag{38}$$

where $\mathbf{c} \in \Re^n$ represents state variable, $\mathbf{H} \in \Re^{n\times n}$, $\mathbf{o} \in \Re^n$.

Theorem 1. *Assuming that* $\mathbf{o}$ *is bounded, if* $\mathbf{H}$ *is the Hurwitz matrix, then the state variable* $\mathbf{c}$ *is bounded stable.*

The proof of Theorem 1 is given in Appendix A.

According to Theorem 1, in Equation (30), h is bounded, $\tilde{\mathbf{A}}$ is Hurwitz, and $\tilde{\mathbf{x}}$ is bounded stable, that is, LESO is bounded stable.

The system model of the MJ-AUV is

$$\dot{z} = \widehat{A}z + \widehat{B}\widehat{u} + f, \tag{39}$$

with $f = \begin{bmatrix} 0 & f_\theta & 0 & f_\psi \end{bmatrix}^{\mathrm{T}}$. Because $\widehat{u} = \widehat{u}_{\mathrm{f}} + \widehat{u}_0$. Because, Equation (39) can be expressed as:

$$\dot{z} = \widehat{A}z + \widehat{B}\left(\widehat{u}_{\mathrm{f}} + \widehat{u}_0\right) + f. \tag{40}$$

By substituting Equations (32) and (34) into Equation (40), we can obtain

$$\dot{z} = \left(\widehat{A} - \widehat{B}K\right)z + \tilde{f}, \tag{41}$$

with $\tilde{f} = f - \hat{f}$. In addition, substituting Equation (37) into Equation (41) obtains

$$\dot{z} = \left(\widehat{A} - \widehat{B}R^{-1}B^{\mathrm{T}}P\right)z + \tilde{f}. \tag{42}$$

$\tilde{f}$ is bounded according to LESO stability analysis, and $\widehat{A} - \widehat{B}R^{-1}B^{\mathrm{T}}P$ is Hurwitz. According to Theorem 1, z is bounded stable.

4. Simulation and Results

In this part, the dynamic model of the MJ-AUV was built on the SIMULINK platform, and the fourth-order Runge–Kuta algorithm was used to solve the second-order differential dynamics equation. Then, the various motions of the model were simulated and analyzed, and the attitude control algorithm of MJ-AUV was simulated and verified. Table A1 shows the model parameters of the MJ-AUV.

4.1. Model Simulation

Figure 5 shows the simulation results of MJ-AUV's pitching motion, and the pitching angle of joint 1 is set to $\theta_1 = 0°$, the yaw angle of joint 1 is set to $\psi_1 = 0°$, the pitch angle of joint 2 rotates in a sinusoidal manner with an amplitude of 20° and a period of 50 s, that $\theta_2 = 20° \sin(\frac{\pi t}{25})$, the yaw angle of joint 2 is set to $\psi_2 = 0°$, and the thrust of the propeller is set to $T = 60$ N. The simulation duration is 100 s. It can be seen from Figure 5a,e that the vehicle has a tendency to move upwards, which is caused by the buoyancy force of the vehicle being greater than its gravity. During the pitching motion, the results in Figure 5a,c show that the vehicle has no change in displacement and velocity in the Y direction, while the results in Figure 5b,d show that the attitudes and angular velocities of the vehicle have no changes in the yaw and roll directions. Therefore, if the vehicle is only controlled in pitching motion, the relative state quantity of the three degrees of freedom can be zero, thus simplifying the complexity of the model.

Figure 6 shows the simulation results of the sinuous motion of the MJ-AUV. The pitching angle of joint 1 is set to $\theta_1 = 0°$, the yaw angle of joint 1 rotates in a sinusoidal manner with an amplitude of 10° and a period of 50 s, i.e., $\psi_1 = 10° \sin(\frac{\pi t}{25})$, the pitch and yaw angle of joint 2 is set to 0°, and the thrust of the propeller is set to $T = 60$ N. The simulation duration is 100 s. In the process of this movement, it can be seen from Figure 6b,d that the MJ-AUV is in roll motion. This is because when the yaw angle of the joint changes, the metacenter of the diversion cabin is not in the same vertical plane as the metacenter of the other cabins, so the roll moment is generated, and then the rolling phenomenon occurs.

Figure 7 shows the result of the spiral diving motion of the MJ-AUV. The pitching angle of joint 1 is set to $\theta_1 = 0°$, the yaw angle of joint 1 is set to $\psi_1 = 5°$, the pitch angle of joint 2 is set to $\theta_2 = 10°$, the pitch angle of joint 2 is set to, the yaw angle of joint 2 is set to $\psi_2 = 0°$, and the thrust of the propeller is set to $T = 60$ N. The simulation duration is 1000 s. Unfortunately, it can be found from the results in Figure 7b,d that the rolling attitude of the vehicle is unstable, which is the same as that of the vehicle in the sinuous motion in Figure 7. It is all caused by the rolling moment that the metacenter of the diversion cabin is not in the same vertical plane as the metacenter of the other cabins. This brings a big problem for the multi-degree of freedom control.

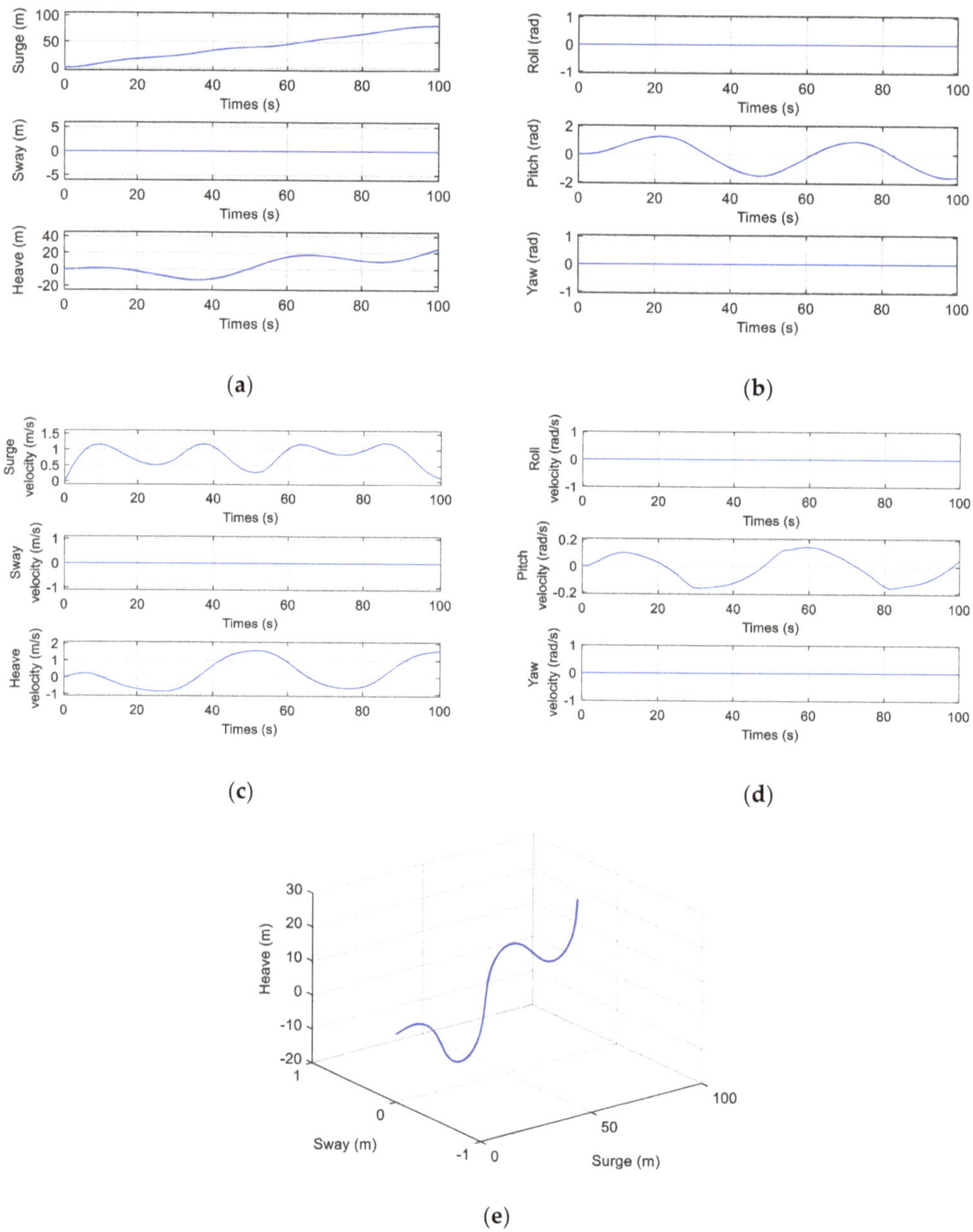

Figure 5. Pitching motion of MJ-AUV: (**a**) Position changes over time; (**b**) Attitude changes over time; (**c**) Linear velocity changes over time; (**d**) Angular velocity changes over time; (**e**) 3D motion.

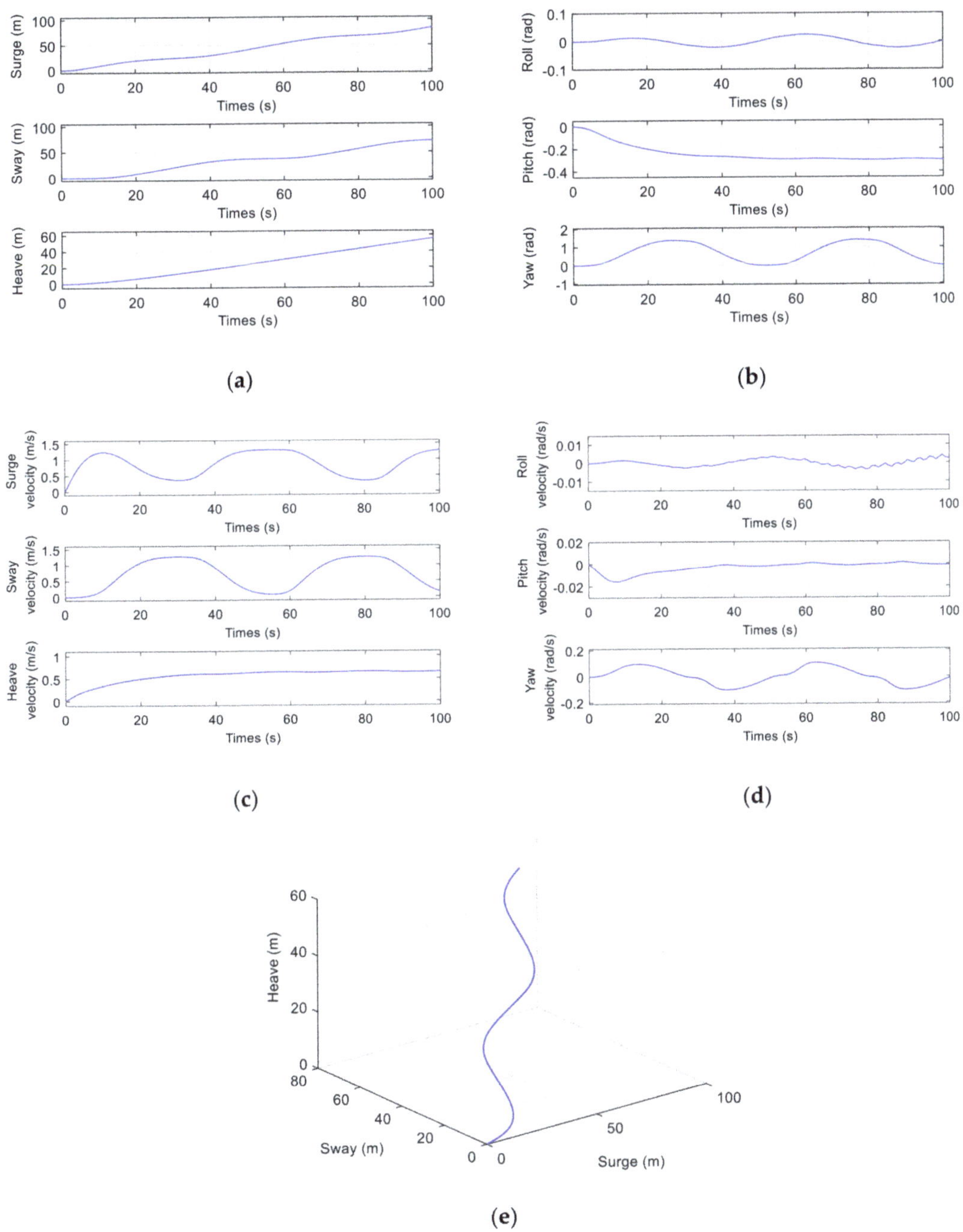

Figure 6. Yaw motion of the MJ-AUV: (**a**) Position changes over time; (**b**) Attitude changes over time; (**c**) Linear velocity changes over time; (**d**) Angular velocity changes over time; (**e**) 3D motion.

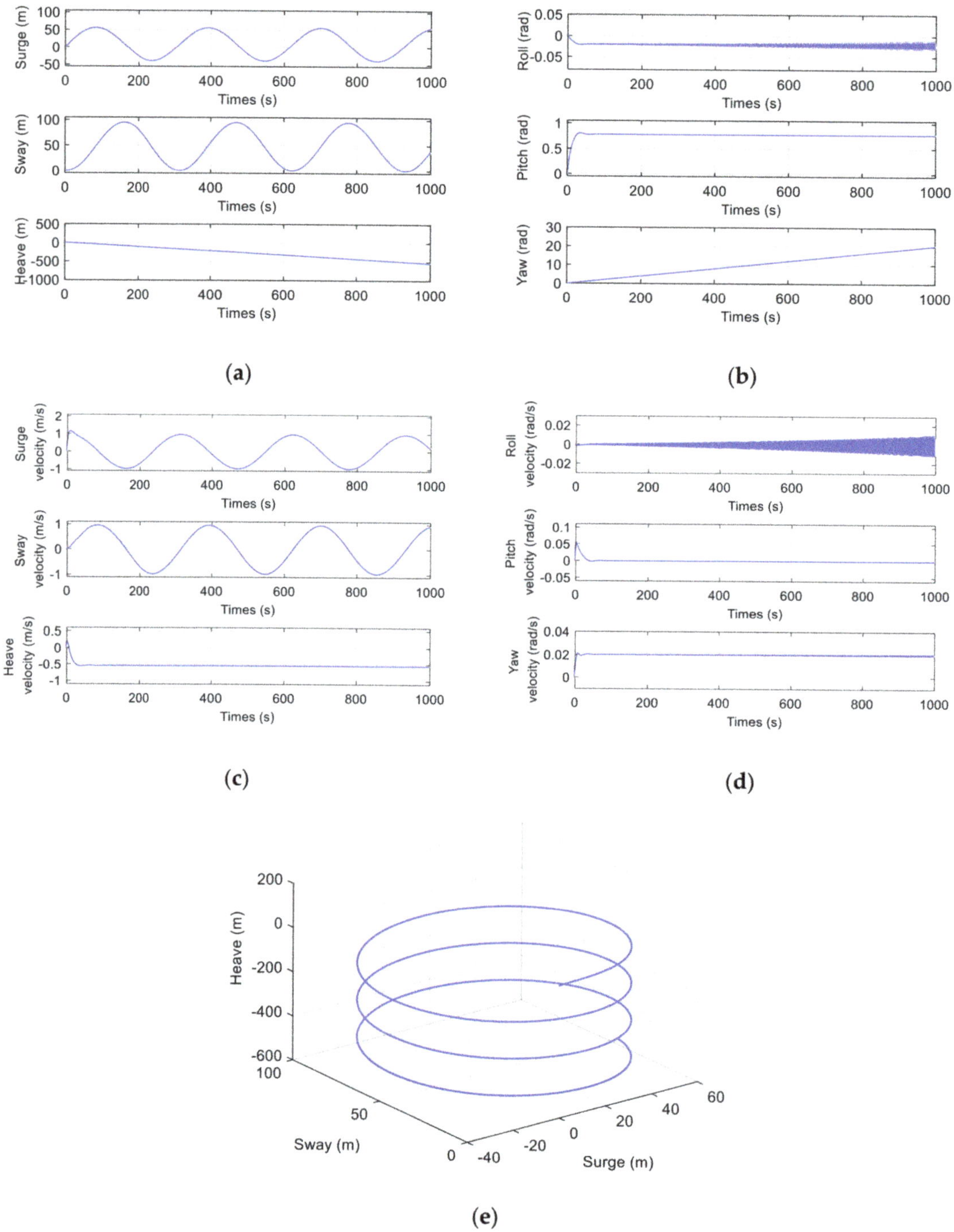

Figure 7. Spiral dive motion of the MJ-AUV: (**a**) Position changes over time; (**b**) Attitude changes over time; (**c**) Linear velocity changes over time; (**d**) Angular velocity changes over time; (**e**) 3D motion.

The above simulation results are consistent with the common physical phenomena, and to a certain extent, it can be considered that the established MJ-AUV 3D dynamic model is reasonable, without loss of generality and reliability. According to the simulation results, compared with the single rigid body AUV, the MJ-AUV has stronger flexible maneuvering characteristics and can perform more complex movements, such as large angle pitching movement, small radius steering movement, spiral ascending and diving movement, and so on, which satisfies the design intention.

4.2. Model and Control Parameters

The initial states of the MJ-AUV are $\theta = 0$ rad, $\dot{\theta} = 0$ rad, $\theta_2 = 0$ rad, $\psi = 0$ rad, $\dot{\psi} = 0$ rad, and $\psi_1 = 0$ rad. In addition, the controller parameters are $\omega_0 = 30$, $b_{11} = 0$, $b_{12} = 0.1$, $b_{21} = 0.1$, $b_{22} = 0$, $\boldsymbol{Q} = \mathrm{diag}(500, 600, 500, 600)$, $\boldsymbol{R} = \mathrm{diag}(0.1, 0.1)$, and diag represents the diagonal matrix. The control gain after LQR optimization is

$$\boldsymbol{K} = \begin{bmatrix} 0 & 0 & -70.7107 & 78.3672 \\ -70.7107 & 78.3672 & 0 & 0 \end{bmatrix}.$$

4.3. Control Simulation

Figures 8–12 show the simulation results of the pitch and yaw attitude control of the MJ-AUV, that is, when the pitch attitude is kept as 0 rad, the yaw attitude is adjusted according to the design signal. Figure 8 shows the pitch control results. The reason why the initial pitch attitude is not 0 rad is that the rest buoyancy of each cabin of MJ-AUV is positive, and the resultant moment of the body is not 0 rad. In other words, when the joint angle is 0 rad, the pitch attitude of the body is not 0 rad, so it is necessary to adjust the body attitude to 0 rad by adjusting the joint angle, which can also be seen from the performance of joint 2 in Figure 11. Figure 9 and the performance of joint 1 in Figure 11 show that, the yaw subsystem can be adjusted quickly after encountering multiple step signals, and the overshoot quantity is nearly 0 rad, with a very smooth transition process. Figure 10 shows that the maximum control error is within 0.005%, verifying that the designed controller has a high control accuracy. Figure 12 shows the observer's estimation results for the total disturbance of the pitching and yaw subsystem. The results show that the estimated value of the total disturbance can follow the real value in real time and accurately, further proving the effectiveness of LESO and making a great contribution to the control of the disturbance compensation.

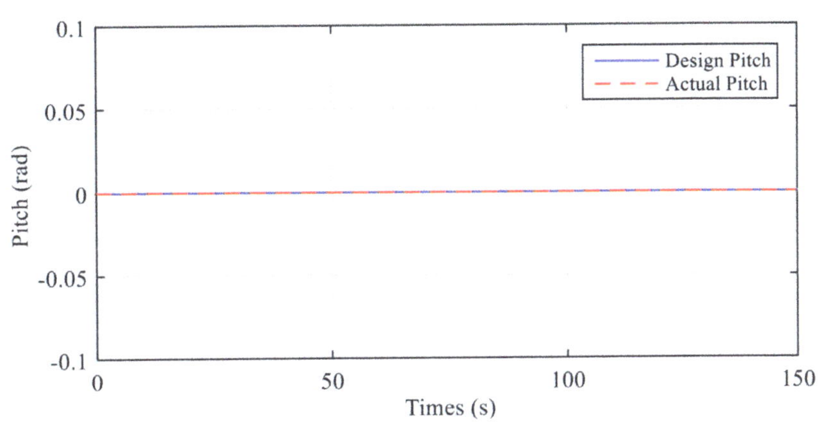

Figure 8. The pitching attitude of the MJ-AUV.

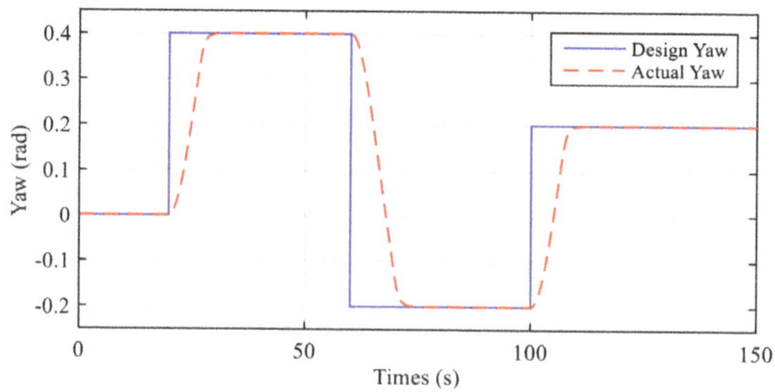

Figure 9. The yaw attitude of the MJ-AUV.

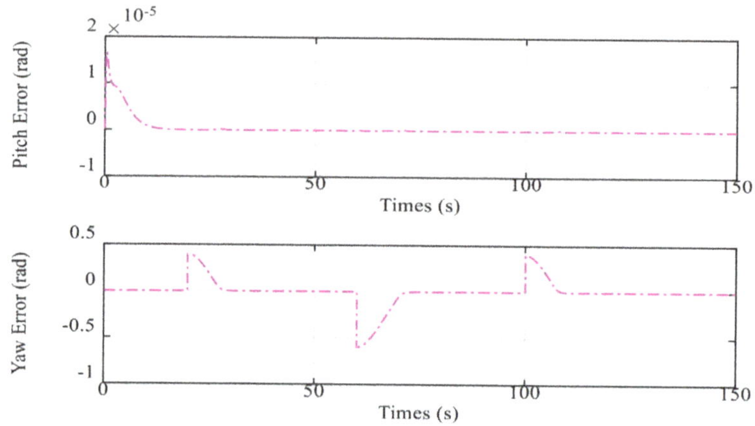

Figure 10. The feedback error of the MJ-AUV.

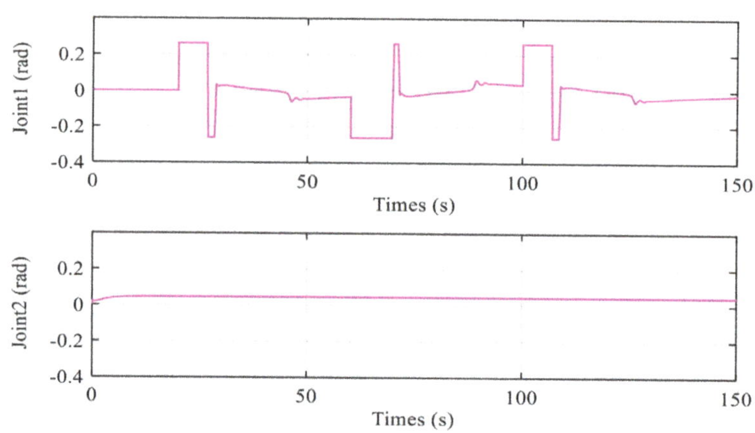

Figure 11. The input of the MJ-AUV.

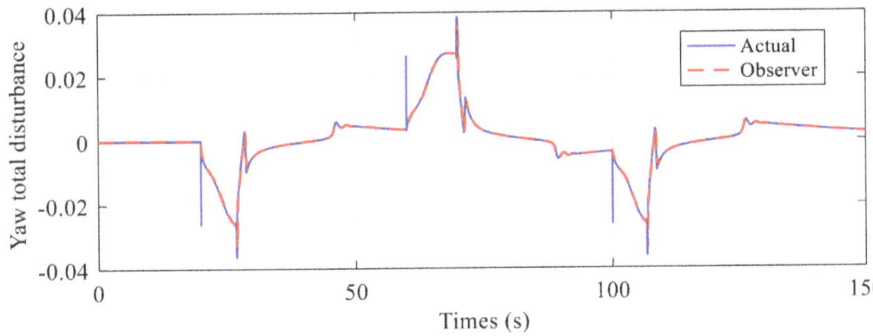

Figure 12. The estimation of the total disturbances of the MJ-AUV.

5. Conclusions

The main contributions of this paper are as follows:

(1) A new type of MJ-AUV with orthogonal joints structure is proposed that can be used for the rapid exploration of the vertical profile of deep sea.
(2) Considering the imbalance of gravity and buoyancy, a 3D analytical model of the MJ-AUV is established according to Newton's second law and torque balance principle, which provides a theoretical basis for the study of the MJ-AUV's 3D motion and model-based controller design.
(3) The LQR decoupling control method based on LESO is proposed. The system is decoupled into pitch and yaw subsystems, and uses linear state observer to estimate and compensate for the total disturbance of each subsystem. LQR is used to achieve the optimal linear feedback control gain according to the expected input and output effects. It solves the problem of strong system coupling and makes the parameter tuning work more efficient.
(4) The simulation results show that the improved control algorithm has the advantages of low overshot and high control precision, and the controller has the advantages of small computation, independence of the precise model of the system, and has a good prospect of engineering application.

Author Contributions: Conceptualization, Q.M. and H.Z.; methodology, L.Y.; software, L.Y.; validation, L.Y.; formal analysis, L.Y.; writing—original draft preparation, L.Y.; writing—review and editing, Q.M., H.Z. and L.Y. All authors have read and agreed to the published version of the manuscript.

Funding: This research was supported by National Key R&D Program of China under Grant No. 2017YFC0306200.

Institutional Review Board Statement: Not applicable.

Informed Consent Statement: Not applicable.

Data Availability Statement: Data sharing not applicable.

Acknowledgments: The authors would like to express their appreciation to Liu Kexian for providing the hydrodynamic coefficient of the vehicle and assisting to complete the modeling work. At the same time, the authors are very grateful to Wang Runzhi for help in the controller design process.

Conflicts of Interest: The authors declare no conflict of interest.

Appendix A

Proof of Theorem 1. The Lyapunov function is defined as

$$V = c^{\mathrm{T}} P c, \tag{A1}$$

where, P is the unique solution of the Lyapunov equation $H^{T}P + PH = -Q$, H is the Hurwitz matrix, and Q is the positive definite matrix, then the derivative of Equation (A1) is obtained as follows:

$$\begin{aligned}\dot{V} &= -c^{T}Qc + 2o^{T}Pc \\ &= \left(c^{T}Q^{\frac{1}{2}} - o^{T}PQ^{-\frac{1}{2}}\right)\left(c^{T}Q^{\frac{1}{2}} - o^{T}PQ^{-\frac{1}{2}}\right)^{T} + \left(o^{T}PQ^{-\frac{1}{2}}\right)\left(o^{T}PQ^{-\frac{1}{2}}\right)^{T}\end{aligned} \tag{A2}$$

Let $\dot{V} < 0$, that

$$\left\|c^{T}Q^{\frac{1}{2}} - o^{T}PQ^{-\frac{1}{2}}\right\|_{2} > \left\|o^{T}PQ^{-\frac{1}{2}}\right\|_{2}. \tag{A3}$$

Processing Equation (A3), can be obtained

$$\left\|c^{T}Q^{\frac{1}{2}}\right\|_{2} > 2\left\|o^{T}PQ^{-\frac{1}{2}}\right\|_{2}. \tag{A4}$$

The choice of P depends on Q. In order to make the calculation simple, $Q = I$ is selected as the identity matrix to meet the requirements of positive definite matrix. Then Equation (A4) is:

$$\|c\|_{2} > 2\|Po\|_{2}. \tag{A5}$$

Because o is bounded, when c satisfies Equation (A5), $\dot{V}$ is negative definite, and because V is positive definite, c is bounded stable. □

Appendix B

Table A1. The Parameters of the MJ-AUV Model.

	Diversion Cabin	Navigation/Control Cabin	Propulsion Cabin
l_i/m	0.5625	0.775	0.6625
m_i/kg	46.5	113.6	46
V_i/m^3	0.0471	0.1152	0.0467
J_{xx}/kg·m	1.129	1.938	1.314
J_{yy}/kg·m	7.623	29.281	8.877
J_{zz}/kg·m	7.645	29.447	8.999
c_{Dx}^{2}	10.05	0	22.73
c_{Dy}^{2}	75.53	126.5	177.5
c_{Dz}^{2}	75.53	126.5	159.8
c_{Dx}^{1}	0.3404	0	0.5009
c_{Dy}^{1}	6.272	12.61	4.192
c_{Dz}^{1}	6.272	12.61	7.351
λ_{11}	12.44	17.30	13.40
λ_{22}	57.71	109.13	82.42
λ_{33}	57.71	109.13	87.97
λ_{44}	0	0	0.47
λ_{55}	2.6	16.37	7.85
λ_{66}	2.6	16.37	7.44
Other Parameters	$l_c = 0.01$ m, $\rho_{liquid} = 1000$ kg/m^3, $g = 9.8$ N/kg, $F_t = 60$ N, $-15° \leq \psi_1 \leq 15°$, $-20° < \theta_2 < 20°$.		

References

1. Zereik, E.; Bibuli, M.; Mišković, N.; Ridao, P.; Pascoal, A. Challenges and Future Trends in Marine Robotics. *Annu. Rev. Control* **2018**, *46*, 350–368. [CrossRef]
2. Rudnick, D.L. Ocean Research Enabled by Underwater Gliders. *Annu. Rev. Mar. Sci.* **2016**, *8*, 519–541. [CrossRef] [PubMed]
3. Yang, M.; Yang, S.; Wang, Y.; Liang, Y.; Wang, S.; Zhang, L. Optimization Design of Neutrally Buoyant Hull for Underwater Gliders. *Ocean Eng.* **2020**, *209*, 107512. [CrossRef]

4. Wynn, R.B.; Huvenne, V.A.I.; Le Bas, T.P.; Murton, B.J.; Connelly, D.P.; Bett, B.J.; Ruhl, H.A.; Morris, K.J.; Peakall, J.; Parsons, D.R.; et al. Autonomous Underwater Vehicles (AUVs): Their Past, Present and Future Contributions to the Advancement of Marine Geoscience. *Mar. Geol.* **2014**, *352*, 451–468. [CrossRef]
5. Xia, D. Dynamic Modeling of a Fishlike Robot with Undulatory Motion Based on Kane's Method. *J. Mech. Eng.* **2009**, *45*, 41–49. [CrossRef]
6. Tang, J.; Li, B.; Chang, J.; Wang, C. Design and Dynamic Model of Underwater Gliding Snake-like Robot. *J. Huazhong Univ. Sci. Technol. Nat. Sci.* **2018**, *12*, 89–94. [CrossRef]
7. Kelasidi, E.; Liljeback, P.; Pettersen, K.Y.; Gravdahl, J.T. Innovation in Underwater Robots: Biologically Inspired Swimming Snake Robots. *IEEE Robot. Autom. Mag.* **2016**, *23*, 44–62. [CrossRef]
8. Kelasidi, E.; Pettersen, K.Y.; Gravdahl, J.T.; Liljeback, P. Modeling of Underwater Snake Robots. In Proceedings of the 2014 IEEE International Conference on Robotics and Automation (ICRA), Hong Kong, China, 31 May–7 June 2014; pp. 4540–4547.
9. Kelasidi, E.; Pettersen, K.Y.; Gravdahl, J.T. Modeling of Underwater Snake Robots Moving in a Vertical Plane in 3D. In Proceedings of the 2014 IEEE/RSJ International Conference on Intelligent Robots and Systems, Chicago, IL, USA, 14–18 September 2014; pp. 266–273.
10. Koopaee, M.J.; Pretty, C.; Classens, K.; Chen, X. Dynamical Modeling and Control of Modular Snake Robots with Series Elastic Actuators for Pedal Wave Locomotion on Uneven Terrain. *J. Mech. Des.* **2020**, 142. [CrossRef]
11. Yu, J.; Liu, L.; Tan, M. Dynamic Modeling of Multi-Link Swimming Robot Capable of 3-D Motion. In Proceedings of the 2007 International Conference on Mechatronics and Automation, Harbin, China, 5–8 August 2007; pp. 1322–1327.
12. Kohl, A.M.; Pettersen, K.Y.; Kelasidi, E.; Gravdahl, J.T. Planar Path Following of Underwater Snake Robots in the Presence of Ocean Currents. *IEEE Robot. Autom. Lett.* **2016**, *1*, 383–390. [CrossRef]
13. Kelasidi, E.; Liljeback, P.; Pettersen, K.Y.; Gravdahl, J.T. Integral Line-of-Sight Guidance for Path Following Control of Underwater Snake Robots: Theory and Experiments. *IEEE Trans. Robot.* **2017**, *33*, 610–628. [CrossRef]
14. Fischer, N.; Hughes, D.; Walters, P.; Schwartz, E.M.; Dixon, W.E. Nonlinear RISE-Based Control of an Autonomous Underwater Vehicle. *IEEE Trans. Robot.* **2014**, *30*, 845–852. [CrossRef]
15. Zhao, S.; Yuh, J.; Choi, S.K. Adaptive DOB Control for AUVs. In Proceedings of the IEEE International Conference on Robotics and Automation (ICRA '04), New Orleans, LA, USA, 26 April–1 May 2004; pp. 4899–4904.
16. Wu, N.; Wu, C.; Ge, T.; Yang, D.; Yang, R. Pitch Channel Control of a REMUS AUV with Input Saturation and Coupling Disturbances. *Appl. Sci.* **2018**, *8*, 253. [CrossRef]
17. Wang, T.; Wu, C.; Wang, J.; Ge, T. Modeling and Control of Negative-Buoyancy Tri-Tilt-Rotor Autonomous Underwater Vehicles Based on Immersion and Invariance Methodology. *Appl. Sci.* **2018**, *8*, 1150. [CrossRef]
18. Kang, S.; Rong, Y.; Chou, W. Antidisturbance Control for AUV Trajectory Tracking Based on Fuzzy Adaptive Extended State Observer. *Sensors* **2020**, *20*, 7084. [CrossRef]
19. Yangyang, Z.; Li'e, G.; Weidong, L.; Le, L. Research on Control Method of AUV Terminal Sliding Mode Variable Structure. In Proceedings of the 2017 International Conference on Robotics and Automation Sciences (ICRAS), Hong Kong, China, 26–29 August 2017; pp. 88–93.
20. Rodriguez, J.; Castañeda, H.; Gordillo, J.L. Design of an Adaptive Sliding Mode Control for a Micro-AUV Subject to Water Currents and Parametric Uncertainties. *JMSE* **2019**, *7*, 445. [CrossRef]
21. Huang, Z.; Liu, Y.; Zheng, H.; Wang, S.; Ma, J.; Liu, Y. A Self-Searching Optimal ADRC for the Pitch Angle Control of an Underwater Thermal Glider in the Vertical Plane Motion. *Ocean Eng.* **2018**, *159*, 98–111. [CrossRef]
22. Zhang, Y.; Deng, H.; Li, Y. Depth Control of AUV Using Sliding Mode Active Disturbance Rejection Control. In Proceedings of the 2018 3rd International Conference on Advanced Robotics and Mechatronics (ICARM), Singapore, 18–20 July 2018; pp. 300–305.
23. Shen, Y.; Shao, K.; Ren, W.; Liu, Y. Diving Control of Autonomous Underwater Vehicle Based on Improved Active Disturbance Rejection Control Approach. *Neurocomputing* **2016**, *173*, 1377–1385. [CrossRef]
24. Wang, T.; Wang, J.; Wu, C.; Zhao, M.; Ge, T. Disturbance-Rejection Control for the Hover and Transition Modes of a Negative-Buoyancy Quad Tilt-Rotor Autonomous Underwater Vehicle. *Appl. Sci.* **2018**, *8*, 2459. [CrossRef]
25. Li, H.; He, B.; Yin, Q.; Mu, X.; Zhang, J.; Wan, J.; Wang, D.; Shen, Y. Fuzzy Optimized MFAC Based on ADRC in AUV Heading Control. *Electronics* **2019**, *8*, 608. [CrossRef]
26. Anderlini, E.; Parker, G.G.; Thomas, G. Docking Control of an Autonomous Underwater Vehicle Using Reinforcement Learning. *Appl. Sci.* **2019**, *9*, 3456. [CrossRef]
27. Sun, Y.; Zhang, C.; Zhang, G.; Xu, H.; Ran, X. Three-Dimensional Path Tracking Control of Autonomous Underwater Vehicle Based on Deep Reinforcement Learning. *JMSE* **2019**, *7*, 443. [CrossRef]
28. Sands, T. Development of Deterministic Artificial Intelligence for Unmanned Underwater Vehicles (UUV). *JMSE* **2020**, *8*, 578. [CrossRef]
29. Gao, Z. Scaling and Bandwidth-Parameterization Based Controller Tuning. In Proceedings of the 2003 American Control Conference, Denver, CO, USA, 4–6 June 2003; Volume 6, pp. 4989–4996.
30. Sudano, J.J. A Transformation of Unit Vectors to Simplify Derivations between Earth-Centered and Local North-East-and-up on an Ellipsoid Earth. In Proceedings of the IEEE 1995 National Aerospace and Electronics Conference (NAECON), Dayton, OH, USA, 22–26 May 1995; Volume 2, pp. 745–747.
31. Gao, Z. Active Disturbance Rejection Control: A Paradigm Shift in Feedback Control System Design. In Proceedings of the 2006 American Control Conference, Minneapolis, MN, USA, 14–16 June 2006; pp. 2399–2405.

Journal of
Marine Science and Engineering

MDPI

Article

Predicting Ship Trajectory Based on Neural Networks Using AIS Data

Tamara A. Volkova [1,*], Yulia E. Balykina [2,*] and Alexander Bespalov [1,*]

1 Department of Applied Mathematics, Admiral Makarov State University of Maritime and Inland Shipping, 5/7, Dvinskaya Str., Saint Petersburg 198035, Russia
2 Faculty of Applied Mathematics and Control Processes, Saint Petersburg State University, 7-9, Universitetskaya Naberezhnaya, Saint Petersburg 199034, Russia
* Correspondence: volkovata@gumrf.ru (T.A.V.); j.balykina@spbu.ru (Y.E.B.); bespalov.aleksandr.p@gmail.com (A.B.)

Abstract: To create an autonomously moving vessel, it is necessary to know exactly how to determine the current coordinates of the vessel in the selected coordinate system, determine the actual trajectory of the vessel, estimate the motion trend to predict the current coordinates, and calculate the course correction to return to the line of the specified path. The navigational and hydrographic conditions of navigation on each section of the route determine the requirements for the accuracy of observations and the time spent on locating the vessel. The problem of predicting the trajectory of the vessel's motion in automatic mode is especially important for river vessels or river-sea vessels, predicting the trajectory of the route sections during the maneuvering of the vessel. At the moment, one of the most accurate ways of determining the coordinates of the vessel is by reading the satellite signal. However, when a vessel is near hydraulic structures, problems may arise connected with obtaining a satellite signal due to interference and, therefore, the error in measuring the coordinates of the vessel increases. The likelihood of collisions and various kinds of incidents increases. In such cases, it is possible to correct the trajectory of the movement using an autonomous navigation system. In this work, opportunities of the possible application of artificial neural networks to create such a corrective system using only the coordinates of the ship's position are discussed. It was found that this is possible on sections of the route where the ship does not maneuver.

Keywords: AIS Data; trajectory prediction; waterway transportation; neural networks; autonomous navigation

Citation: Volkova, T.A.; Balykina, Y.E.; Bespalov, A. Predicting Ship Trajectory Based on Neural Networks Using AIS Data. *J. Mar. Sci. Eng.* **2021**, *9*, 254. https://doi.org/10.3390/jmse9030254

Academic Editors: Evgeny Veremey and Carlos Guedes Soares

Received: 19 January 2021
Accepted: 22 February 2021
Published: 28 February 2021

Publisher's Note: MDPI stays neutral with regard to jurisdictional claims in published maps and institutional affiliations.

1. Introduction

Nowadays, to build the trajectory of the vessel and determine its position, it is a common approach to use data from satellite navigation systems and automatic identification systems (AIS), which allow obtaining coordinates with high accuracy and secondly, discreteness. However, despite the advances in the development of satellite navigation technologies in recent decades, their methods have a significant drawback: non-autonomy, which does not allow the use of a satellite system when a number of known circumstances occur. The main ones are the loss of communication or, under the conditions of maneuvering, too long processing the received data, which prevent it from making a decision in time. In this regard, the issues of studying autonomous methods for determining a ship's position are of great practical interest. The prediction of the trajectory of the vessel's movement is of practical importance for the development and creation of unmanned autonomous vessels.

Taking into account the inertia of the vessel, in order to react in time to possible obstacles, it is necessary to predict in advance the estimated nearest coordinates of the vessel. In this regard, it is necessary to take into account the capabilities of the ship's technical means, so that the time required for the autopilot to make a decision is sufficient enough to implement it. In this sense, predicting the trajectory of a vessel's movement based on historical data, that is, using data received from other vessels, can only be used very approximately, since the trajectory of a vessel's movement depends quite strongly on specific external conditions. Therefore, the goal of this work was to analyze the application of neural network technologies that use knowledge of the previous coordinates of the vessel's trajectory to predict the estimated next coordinates of the vessel during river navigation.

Related Works

Various studies have examined different aspects of predicting vessel trajectories using AIS data, including behavior anomaly detection [1], waypoint detection [2], and the actual forecast of the vessel's trajectory [3–5]. Since artificial neural networks gained wide popularity, a number of studies have appeared that focused on the use of artificial intelligence in solving problems related to navigation. For example, there is a significant difference between the different patterns of vessel traffic, namely static navigation, normal navigation, and maneuvering. As the data that are collected during maneuvering are of the greatest interest, a separate issue is the task of separating them from the general data. In this regard, there are a number of studies that use artificial neural networks to classify AIS data types in relation to ship traffic [6].

Neural networks are often used to study the trajectories of various types of ships in the oceans [7–9]. They allow one to study the possible routes of a ship's motion, and identify the most frequently used ones. There are various probabilistic approaches for solving this kind of problem as well [10]. There are studies that, while solving this problem, also touch upon the problem of using neural technologies to predict possible ship collisions [11]. It is believed that the identification and construction of optimal routes will allow the use of unmanned vessels, which will follow the given trajectories [12–14]. When using neural networks to predict the motion of the vessel and build the trajectory of its motion, neural networks are used either as a function for identifying the model of the vessel movement, or for the trajectory prediction, and the correction of their free parameters is performed in real time as the vessel is moving [15–17].

A detailed review of various approaches to predicting the trajectory of vessel movement or location is given in [18]. Both classical models, such as Constant Velocity Model (CVM), and modern approaches based on machine learning and statistical analysis are analyzed. Models for predicting the trajectory of a specific vessel based on neural networks using course of ground (COG) and speed over ground (SOG), attached with a corresponding future timestamp as input, are discussed in [19], and based on the obtained data the coordinates of the ship are restored.

2. Materials and Methods

Unlike the movement of sea vessels, which are faced with maneuvering in confined spaces mainly when passing straits and moving in port waters, river vessels are forced to constantly maneuver in confined spaces (e.g., the presence of a large number of possible obstacles, limited fairway, frequent course changes, a large number of oncoming and passing vessels). Moreover, the performance of satellite navigation systems is greatly influenced by the presence of possible sources of industrial interference, as well as a large number of hydraulic structures, located on the riverbanks. Therefore, AIS data are not error-free. Different sources of error that render a portion of these data noisy and inadequate for monitoring are discussed in detail in [20].

Since the aim of the research was to determine the possibility of predicting the position of the vessel at the next moment in time using only the previous coordinates of the vessel's movement for training the neural network, only these data were collected during the experiment. The real experiment was carried out specifically for collecting data in the conditions of river navigation, as the data in existing open databases often consist of information collected during sea navigation.

It was assumed that the problem of predicting the vessel trajectory could be solved on the basis of a focused time lagged feedforward network. For training the neural network model, the data obtained during the experiment were used. The experiment was carried out on a ship with characteristics presented in Table 1:

Table 1. Vessel characteristics.

Feature	Value	Units
Maximum length	35.65	meters
Maximum width	5.80	meters
Side height	2.60	meters
Draft	1.51	meters
Displacement	157	tons
Main engine power	2 × 200	kW
Crew	10	man
Freeboard	1.096	meters
Endurance	6	days

There were two antennas placed on the roof of the ship's cabin at a distance of 90 cm along the center plane of the ship and 66 centimeters between the antennas. The data from the second antenna underwent a post-processing and were subsequently received as a reference. All vessel position data were recorded using the NMEA-0183 protocol.

NMEA is a format for transmitting data between ship instruments. It includes a message system for information exchange between GPS navigation receivers and navigation information consumers. All commands and messages are transmitted in ASCII text form. The last field of the message may contain the checksum of the current message, starting with the * separator. This is followed by an 8-bit checksum (exclusive OR) of all characters in the message, including spaces between the $ and * separators, not including the last one. The hexadecimal result is translated into two ASCII characters (0–9, A–F).

The GGA-GPS data, which were obtained during the experiment and used to train neural networks and verify the results, contained the following values:

- Greenwich Mean Time at the moment of determining the location.
- Latitude.
- North/South (N/S).
- Longitude.
- West/East (E/W).
- GPS signal quality indicator:
  0 = Positioning is not possible or is not correct;
  1 = GPS mode, normal accuracy, location possible;
  2 = Differential GPS mode, normal accuracy, location possible;
  3 = GPS precision mode, location possible.
- The number of satellites used (0–12, may differ from the number of visible ones).
- Horizontal Dilution of Precision (HDOP).
- Receiver antenna height above/below sea level.
- Unit of antenna's location height, meters.
- Geoid difference, i.e., the difference between the WGS-84 ellipsoid and the sea level (geoid), "-" = sea level is below the ellipsoid.
- Units of measurement, meters.
- GPS Differential Data Age—Time in seconds since the last SC104 type 1 or 9 update, filled with zeroes if the differential mode is not used.
- ID of the station transmitting differential corrections, ID, 0000-1023.
- Checksum.

An example of the received data is presented in Figure 1.

```
$PNVGDBG,085228.20,1,6.8,38,0.0016576,-0.0076*62
$PNVGIMU,085228.20,-4.60,3.86,320.70,0.07,-0.22,-0.03*7B
$PNVGIRQ,28272,28272,0,2828960,2828960,0,2828960,2828960,0*59
$PNVGCPU,38774,33463,0,2,0,17617,9177,0,0*67
$PNVGKF0,332,7,436,86,407,701,101,1128,60,24,26*1F
$PNVGKF1,228,4,139,57,463,539,60,481,74,24,13*18
$GPRMC,085228.40,A,5952.0525254,N,03058.2179220,E,3.99,312.71,270918,,,R*45
$GPGGA,085228.40,5952.0525254,N,03058.2179220,E,4,15,0.8,31.883,M,15.787,M,0.4,0000*7A
$GPGSV,5,1,18,7,4,308,40,8,41,296,46,10,56,180,51,13,15,23,41*77
$GPGSV,5,2,18,15,30,61,51,16,22,216,46,18,6,268,38,20,67,117,49*72
```

Figure 1. Sample data received from the first antenna.

To obtain images correlated with geographic maps, the algorithm for transferring data from the NMEA protocol to the WGS84 system (World Geodetic System 1984) was used [21]:

$$\text{flTemp} = \text{Value (NMEA 0183)}$$

$$\text{nTemp} = \text{flTemp}/100.0;$$

$$\text{nTemp} = \text{nTemp} - (\text{nTemp}\%\ 1);$$

$$\text{flMin} = \text{flTemp} - 100.0 * \text{nTemp};$$

$$\text{Lo} = \text{nTemp} + \text{flMin}/60.0;$$

where /—division, %—remainder of division, *—multiplication, Lo—value in WGS84.

3. Results and Discussion

The experiment was conducted on a segment of inland waterways of the Neva-Ladoga region. The resulting vessel path is shown in Figure 2. It should be noted that the results of all antennas are very close to each other.

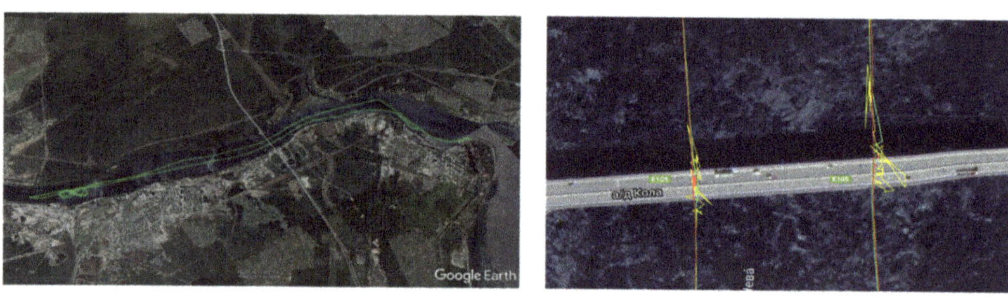

(**a**) (**b**)

Figure 2. Total path of the ship (**a**), the part of the path near and under the bridge (**b**).

During the experiment, the case of a ship passing under a bridge was considered separately. Since the passage of the vessel under the bridge is quite difficult due to the small distance between the bridge supports, the flow of the river, etc., in order to reduce the risk of a dangerous situation, it is necessary to have alternative methods of determining the exact location of the vessel.

One can see (Figures 3–5) how the received signals behave while passing the bridge. It is clear that the signal is either interrupted, or does not reflect in any way the real trajectory of the vessel. This fact is most evident for the processed signal when the system is trying to restore the trajectory.

In this regard, the task was set to assess the possibilities of using neural networks to solve such problems, in particular, to determine whether it is possible to build a neural network that will restore the trajectory of a ship using the previous AIS data. Neural networks with various activation functions were studied.

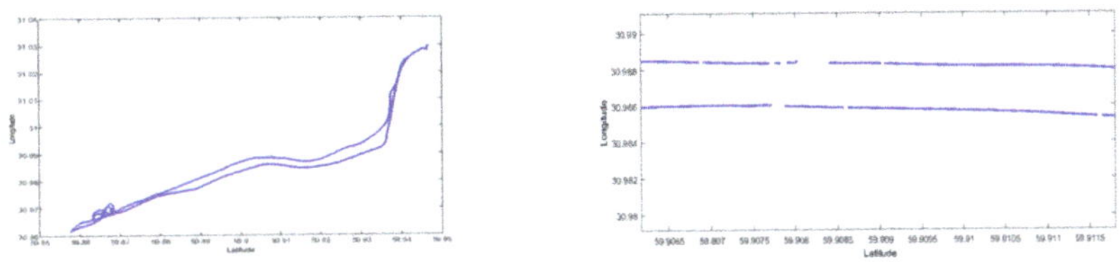

(**a**) (**b**)

Figure 3. Vessel trajectory, obtained from the first antenna (**a**), vessel trajectory under the bridge, obtained from the first antenna (**b**).

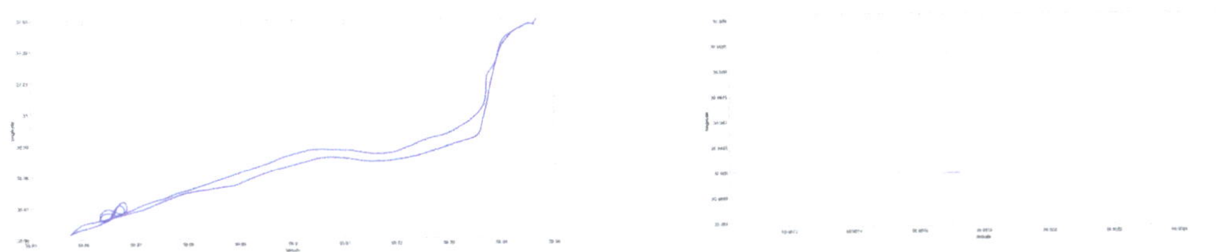

(**a**) (**b**)

Figure 4. Vessel trajectory, obtained from the second antenna (**a**), vessel trajectory under the bridge, obtained from the second antenna (**b**).

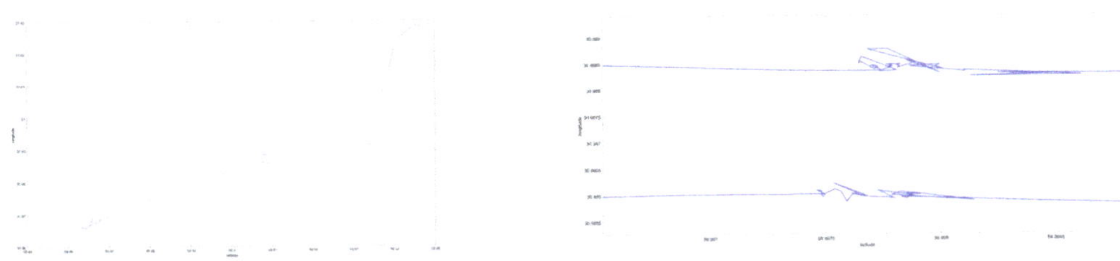

(**a**) (**b**)

Figure 5. Vessel trajectory (**a**); trajectory under the bridge, the result of post-processing the signal (**b**).

To train the neural network, an array of 1000 lines containing latitude and longitude was taken (see Table 2 for data sample).

Table 2. Input data sample.

Latitude	Longitude
59.9466765683333	31.0296530033333
59.9466765750000	31.0296530366667
59.9466765833333	31.0296530600000
59.9466765916667	31.0296530766667
59.9466765900000	31.0296531200000
59.9466765883333	31.0296531333333
59.9466765733333	31.0296531333333

There was 10% of the initial set left for the validation process. Of the remaining 90%, 70% was taken as a training set, and 20% was left for testing.

The constructed neural network consisted of two layers, the first one contained neurons with sigmoidal activation functions, and the other one had one neuron with identical activation functions. The network, taking the vector of the delayed values of the ship's coordinates in the geographic system as an input, predicts its value one step ahead. The network is trained in real time based on a sample containing a certain number of previous trajectory points.

The Levenberg-Marquardt algorithm, which implements the idea of gradient descent, was used as a training algorithm. The mean square of the course error was chosen as a measure of the training accuracy. This algorithm most adequately reproduces the dynamics of the vessel's movement. Its advantages, in comparison with other methods of conjugate gradients, are high speed calculations and ensuring convergence.

The first of the constructed networks included 200 input neurons. The hyperbolic tangent was used as a neuron activation function (1).

$$a = \frac{2}{(1 + \exp(-2 \cdot n))} - 1 \tag{1}$$

The network with 200 neurons in the hidden layer was trained in 50 iterations, and the validation showed the difference in latitude by 10^{-6}, which is an insignificant discrepancy from the true data, since the numbers after 6 decimal places are not significant (Table 3). Resulting average error on the straight sections of the vessel path was 0.02° An increase in the number of neurons did not lead to a decrease in errors, but increased the calculation time.

Table 3. Results of the first neural network.

	Input Data	Output Data	Reference Data
Latitude	59.8665262166667	59.866514809161487	59.8665157383333
Longitude	30.970706225	30.969370218195859	30.9706893433333

The results of the first neural network performance are presented in Figures 6 and 7.

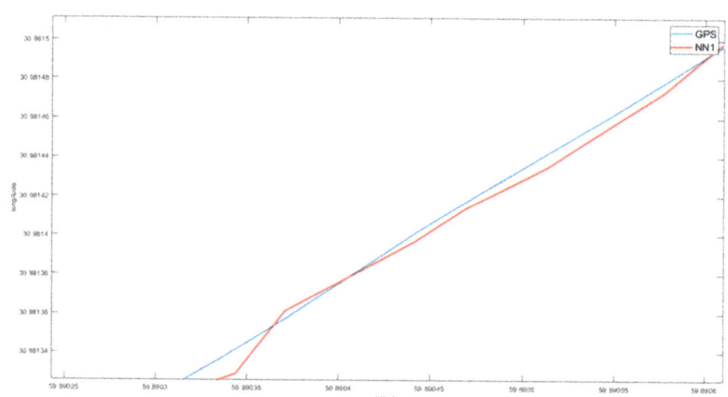

Figure 6. Comparison of the trajectory predicted by the neural network with the reference trajectory on the straight section of the path.

The second of the options for building and training a neural network was a network with the following structure: the first layer was the input data for the neural network and it did not participate in the calculation; the second layer was a hidden layer; and the third one processed the data received from the second layer and returned the desired value. For the convenience of working with the neural network, the input data were transformed as follows: the GPS coordinate values were divided by 100 to bring them to the interval (0; 1).

Subsequently, the data were returned to their original state. To train the neural network, 24,000 datasets were used with 10,000 iterations for each dataset.

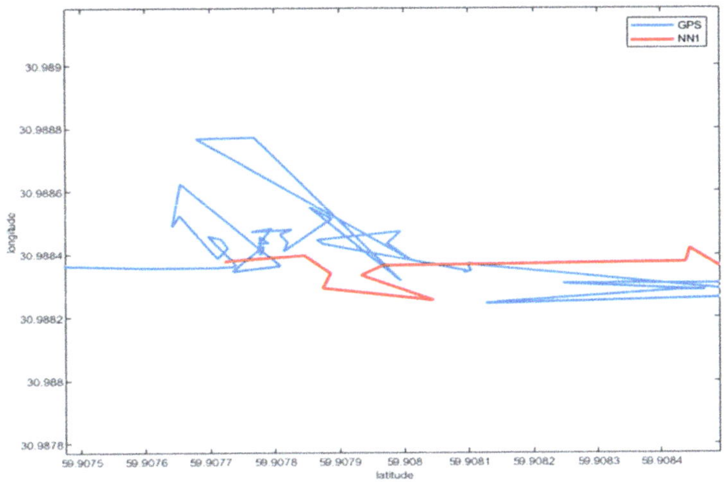

Figure 7. Comparison of the trajectory predicted by the neural network with the reference trajectory along the path with data loss and distortion.

Function (2) was used as an activation function.

$$\sigma(x) = \frac{1}{(1 + \exp(-tx))} \tag{2}$$

In this case, t is the weight of the neuron, which is automatically selected during training of the neural network. The peculiarity of neurons with such a transfer characteristic is that they amplify strong signals much less than weak ones, since the areas of strong signals correspond to the flat areas of the activation function.

It should be noted that functions (1) and (2) differ only in the range of values, however, this difference significantly changes the possible input and output values of the neuron, which subsequently affects the learning rate and accuracy of the neural network.

For the input data, 10 pairs of latitude and longitude values were used, from which the values of the next coordinate were obtained. Data in Table 4 show that the accuracy of the data produced by the neural network differs for the worse, since in this case the difference in latitude from the true data is 10^{-5}.

Table 4. Results of the second neural network.

Input Data		Output Data	
59.8902644083333	30.9812980466667	59.8903520000000	30.9813470000000
59.8902644083333	30.9812980466667		
59.8902644083333	30.9812980466667		
59.8903000583333	30.9812980466667		
59.8903000583333	30.9813171183333		
59.8903000583333	30.9813171183333		
59.8903000583333	30.9813171183333		
59.8903000583333	30.9813171183333		
59.8903360133333	30.9813171183333		
59.8903060000000	30.9813420000000		

The results of the second neural network are presented in Figures 8 and 9.

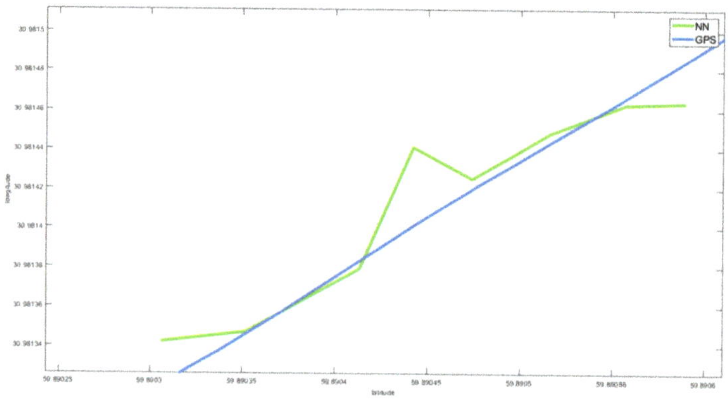

Figure 8. Comparison of the trajectory predicted by the neural network with the reference trajectory on the straight section of the path.

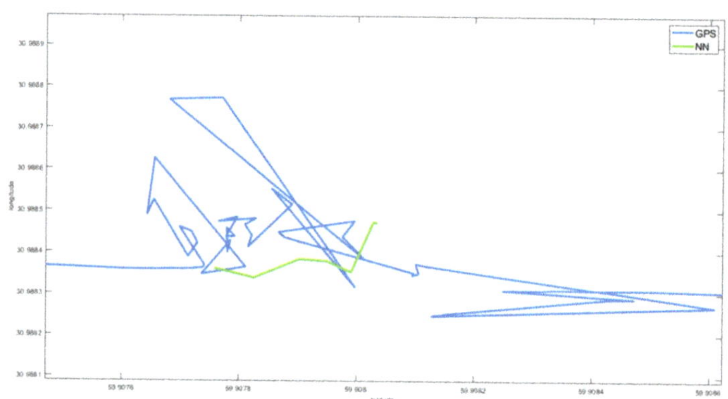

Figure 9. Comparison of the trajectory predicted by the neural network with the reference trajectory along the path with data loss and distortion.

While assessing the performance of the second neural network, the following peculiarities appear:

- the considered neural network can predict the trajectory of a vessel's motion on straight sections; however, due to the accuracy of such a prediction, it is impossible to guarantee the safety of the vessel's motion (Figure 8);
- in areas with GPS data loss or distortion, the first few iterations of the neural network show a good result, which can be seen in Figure 9;
- the developed neural network is not suitable for long-term use due to the lack of stable prediction quality.

Figures 10 and 11 show the comparison of the results of the neural networks' performance with respect to each other and the trajectory obtained from the reference coordinates.

Despite the fact that the second network obviously gives a better result than the first one, it is still far from the true trajectory of the ship. In the case when the ship begins to make any maneuvers (e.g., turn), or enters the interference zone, the available AIS data are insufficient to train neural networks that predict the position or trajectory of the vessel. Therefore, additional research is needed to adapt neural network technologies in relation to the physical characteristics of each specific vessel, as well as swimming conditions, characterized by a change in the density of water and air, the influence of shallow water, and other similar effects.

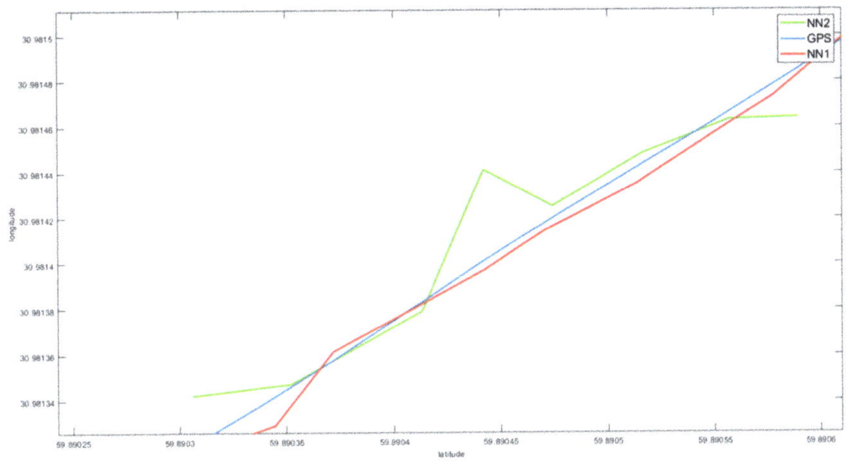

Figure 10. Comparison of the trajectories predicted by the considered neural networks with the reference trajectory on the straight section of the path.

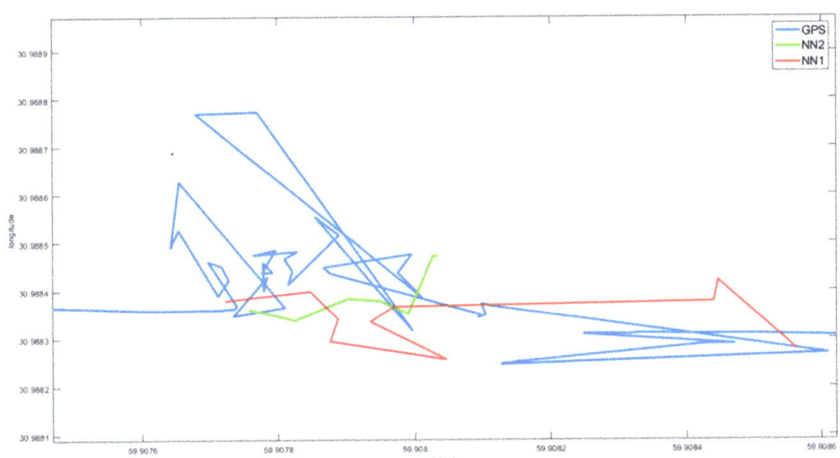

Figure 11. Comparison of the trajectories predicted by considered neural networks with the reference trajectory along the path with data loss and distortion.

4. Conclusions

The problem of determining the position of the vessel in real time is very important for various kinds of applications, in particular, for the autopilot system development. The use of existing algorithms based on AIS data does not allow assessing the position of the vessel in the presence of interference that affects the receipt of AIS signal. One of the ways to solve this issue is the development of an autonomous system that predicts the trajectory of the vessel's movement, which will be a source of additional information for timely decision-making in case of AIS signal distortion. In this paper, the possibility of using neural network technologies is being considered as one of the options for predicting the trajectory of a ship. Neural networks with different activation functions and different structures were built and trained, using experimental data of the real movement of the vessel in the fairway of the Neva River. The goal was to assess whether it was possible to obtain reliable results using only the coordinates of the ship's movement obtained from AIS for neural network training.

Simulation modeling has shown that when the vessel moves along a straight trajectory, the use of neural networks gives quite satisfactory results. The main problems appear when the ship starts to maneuver. With this type of movement, the results shown by trained neural networks are not satisfactory, since the discrepancy between the results of their work and real data is very large. In fact, it is possible to predict the trajectory only one coordinate ahead, which in the conditions of ship maneuvering does not permit using this technology to create autopilots.

Thus, additional research is required to identify which parameters should be added to train neural networks to predict the trajectory during ship maneuvering more accurately. Further research will be devoted to these issues.

Author Contributions: Conceptualization, T.A.V., Y.E.B. and A.B.; methodology, T.A.V.; Y.E.B. and A.B.; validation, T.A.V., Y.E.B. and A.B.; writing—original draft preparation, T.A.V., Y.E.B. and A.B.; writing—review and editing, T.A.V., Y.E.B. and A.B.; supervision, T.A.V., Y.E.B. and A.B. All authors have read and agreed to the published version of the manuscript.

Funding: This research received no external funding.

Institutional Review Board Statement: Not applicable.

Informed Consent Statement: Not applicable.

Data Availability Statement: The data presented in this study are available on request from the corresponding author.

Conflicts of Interest: The authors declare no conflict of interest.

References

1. Sidibé, A.; Shu, G. Study of Automatic Anomalous Behaviour Detection Techniques for Maritime Vessels. *J. Navig.* **2017**, *70*, 847–858. [CrossRef]
2. Dobrkovic, A.; Iacob, M.-E.; Hillegersberg, J. Using machine learning for unsupervised maritime waypoint discovery from streaming AIS data. In Proceedings of the 15th International Conference on Knowledge Technologies and Data-driven Business, Graz, Austria, 21–22 October 2015; pp. 1–8. [CrossRef]
3. Mao, S.; Tu, E.; Zhang, G.; Rachmawati, L.; Rajabally, E.; Huang, G.-B. An Automatic Identification System (AIS) Database for Maritime Trajectory Prediction and Data Mining. *arXiv* **2018**, arXiv:1607.03306. Available online: https://arxiv.org/abs/1607.03306 (accessed on 26 December 2020).
4. Alizadeh, D.; Alesheikh, A.; Sharif, M. Prediction of vessels locations and maritime traffic using similarity measurement of trajectory. *Ann. GIS* **2020**. [CrossRef]
5. Alizadeh, D.; Alesheikh, A.; Sharif, M. Vessel Trajectory Prediction Using Historical Automatic Identification System Data. *J. Navig.* **2021**, *74*, 156–174. [CrossRef]
6. Chen, X.; Liu, Y.; Achuthan, K.; Zhang, X. A ship movement classification based on Automatic Identification System (AIS) data using Convolutional Neural Network. *Ocean Eng.* **2020**, *218*, 108182. [CrossRef]
7. De Vries, G.K.D.; Van Someren, M. Machine learning for vessel trajectories using compression, alignments and domain knowledge. *Expert Syst. Appl.* **2012**, *39*, 13426–13439. [CrossRef]
8. Murray, B.; Perera, L.P. A dual linear autoencoder approach for vessel trajectory prediction using historical AIS data. *Ocean Eng.* **2020**, *209*, 107478. [CrossRef]
9. Soldi, G.; Gaglione, D.; Forti, N.; Di Simone, A.; Daffinà, F.C.; Bottini, G.; Quattrociocchi, D.; Millefiori, L.M.; Braca, P.; Carniel, S.; et al. Space-based Global Maritime Surveillance. Part II: Artificial Intelligence and Data Fusion Techniques. *arXiv* **2020**, arXiv:2011.11338.
10. Rong, H.; Teixeira, A.P.; Guedes Soares, C. Data mining approach to shipping route characterization and anomaly detection based on AIS data. *Ocean Eng.* **2020**, *198*, 106936. [CrossRef]
11. Ahn, J.-H.; Rhee, K.; You, Y. A study on the collision avoidance of a ship using neural networks and fuzzy logic. *Appl. Ocean Res.* **2012**, *37*, 162–173. [CrossRef]
12. Gao, M.; Shi, G.-Y. Ship-handling behavior pattern recognition using AIS sub-trajectory clustering analysis based on the T-SNE and spectral clustering algorithms. *Ocean Eng.* **2020**, *205*, 106919. [CrossRef]
13. Xu, H.; Rong, H.; Guedes Soares, C. Use of AIS data for guidance and control of path-following autonomous vessels. *Ocean Eng.* **2019**, *194*, 106635. [CrossRef]

14. Tu, E.; Zhang, G.; Mao, S.; Rachmawati, L.; Huang, G. Modeling Historical AIS Data for Vessel Path Prediction: A Comprehensive Treatment. *arXiv* **2020**, arXiv:2001.01592.
15. Borkowski, P. The Ship Movement Trajectory Prediction Algorithm Using Navigational Data Fusion. *Sensors* **2017**, *17*, 1432. [CrossRef]
16. Tang, H.; Yin, Y.; Shen, H. A model for vessel trajectory prediction based on long short-term memory neural network. *J. Mar. Eng. Technol.* **2019**, 18. [CrossRef]
17. Zheng, H.; Negenborn, R.R.; Lodewijks, G. Trajectory tracking of autonomous vessels using model predictive control. *IFAC Proc. Vol.* **2014**, *47*, 8812–8818. [CrossRef]
18. Xiao, Z.; Fu, X.; Zhang, L.; Goh, R.S.M. Traffic Pattern Mining and Forecasting Technologies in Maritime Traffic Service Networks: A Comprehensive Survey. *IEEE Trans. Intell. Transp. Syst.* **2020**, *21*, 1796–1825. [CrossRef]
19. Xiao, Z.; Fu, X.; Zhang, L.; Zhang, W.; Liu, R.W.; Liu, Z.; Goh, R.S. Big Data Driven Vessel Trajectory and Navigating State Prediction With Adaptive Learning, Motion Modeling and Particle Filtering Techniques. *IEEE Trans. Intell. Transp. Syst.* **2020**. [CrossRef]
20. Patroumpas, K.; Alevizos, E.; Artikis, A.; Vodas, M.; Pelekis, N.; Theodoridis, Y. Online event recognition from moving vessel trajectories. *Geoinformatica* **2017**, *21*, 389–427. [CrossRef]
21. GeoStar Navigation. GEOS Module. NMEA Communications Protocol Ver. 3.0. Available online: http://geostar-navi.com/files/docs/geos3/geos_nmea_protocol_v3_0_eng.pdf (accessed on 25 December 2020).

Journal of
Marine Science and Engineering

MDPI

Article

Path-Following Control Method for Surface Ships Based on a New Guidance Algorithm

Zhanshuo Zhang [1], Yuhan Zhao [1], Guang Zhao [2], Hongbo Wang [1,*] and Yi Zhao [1,*]

[1] State Key Laboratory on Integrated Optoelectronics, College of Electronic Science and Engineering, Jilin University, Changchun 130000, China; zszhang19@mails.jlu.edu.cn (Z.Z.); yuhanz20@mails.jlu.edu.cn (Y.Z.)

[2] Branch of 707 Institute of China State Shipbuilding Corporation Limited (CSSC), Jiujiang 332007, China; zhaoguang2891@gmail.com

* Correspondence: wang_hongbo@jlu.edu.cn (H.W.); yizhao@jlu.edu.cn (Y.Z.); Tel.: +86-0431-85-14-82-42 (H.W.); +86-0431-85-16-82-80 (Y.Z.)

Abstract: A new type of path-following method has been developed to steer marine surface vehicles along desired paths. Path-following is achieved by a new hyperbolic guidance law for straight-line paths and a backstepping control law for curved paths. An optimal controller has been improved for heading control, based on linear quadratic regulator (LQR) theory with nonlinear feedback control techniques. The control algorithm performance is validated by simulation and comparison against the requirements of International Standard IEC62065. Deviations are within the allowable range of the standard. In addition, the experimental results show that the proposed method has higher control accuracy.

Keywords: ship motion control; path-following; guidance algorithm; nonlinear feedback

Citation: Zhang, Z.; Zhao, Y.; Zhao, G.; Wang, H.; Zhao, Y. Path-Following Control Method for Surface Ships Based on a New Guidance Algorithm. *J. Mar. Sci. Eng.* **2021**, *9*, 166. https://doi.org/10.3390/jmse9020166

Academic Editor: Jakub Montewka

Received: 27 December 2020
Accepted: 1 February 2021
Published: 6 February 2021

1. Introduction

Autopilot is the main equipment for ship motion control; it controls the ship's course without the participation of the helmsman [1]. On autopilot, the ship follows the target route automatically, which can effectively reduce ship operating costs [2]. This paper presents a new method for ship track control. The goal of the control system is to follow a preset route with good anti-interference ability [3]. Path-following control systems for marine vehicles are usually constructed as three independent blocks: guidance, navigation and control [4]. Guidance is the action or the system that continuously computes the reference position, velocity and acceleration of the vehicle to be used by the control system [5]. Navigation is the action or the system that directs the vehicle, by which the position, course and traveled distance is determined. The control system determines the necessary forces and moments to achieve a certain objective.

Line of sight (LOS) is the most commonly used guidance algorithm. This algorithm was first applied to track control of surface ships by Mcgookin and Fossen and has been widely studied [6–9]. The advantage of the LOS guidance algorithm is that it controls the ship's motion by imitating the behavior of the helmsman to eliminate track deviation from the planned route. However, there are some defects in the LOS guidance algorithm. When the distance between the next target waypoint on the target route and the current position of the ship is large, a large track deviation will be produced when the ship tracks the planned route through the guidance algorithm, in addition, the convergence speed of the algorithm is also very slow. In response to this problem, an integral LOS (ILOS) has been proposed and extensively analyzed. Fossen and Lekkas [10] proposed an ILOS based on adaptive control theory that can compensate for drift forces effectively.

Stability analysis is an important topic for navigation and control systems used in autonomous marine vehicles. Proportional-integral-derivative (PID) controllers or proportional derivative controllers are usually used in ship autopilot design [11,12]. In recent years, improved PID control algorithms have emerged. For example, steering parameters for normal adaptable PID autopilots have been developed during the last decade [13,14]. Dlabac et al. [15] presented a particle swarm optimization (PSO)-based PID controller for ship course-keeping. Recently, nonlinear controllers for autopilot motion control of marine vessels have been reported. Oh and Sun [16] presented a model predictive control for path following of underactuated surface vessels. Guerrero et al. [17] employed an adaptive high- order sliding mode controller for trajectory tracking of autonomous underwater vehicles. Designs based on neural networks [18,19], pole placement technique (PPT) [20], fuzzy logic [21], extended state observer technique (ESO) [22] and some other methods are also used. Wang et al. [23] presented a heading control algorithm based on an H-infinite optimization algorithm to counteract the influence of waves and ensure that ships can turn steadily in rough seas. Sun et al. [24] proposed a feedback linearization optimal heading control algorithm to effectively control a ship's course. Veremey et al. [25] proposed a new approach for the compensative transformation of reference dynamic output controllers. Du et al. [26,27] used adaptive robust nonlinear control to adjust a ship's course and track. Xiang et al. [28] improved the fuzzy logic method and tested it on underwater ships and marine surface vehicles. Xu et al. [29] proposed a vector field guidance law to follow the ship's trajectory. Although most of these methods have good results, there is no standard to verify the control effect of their experiments and the ship model they use is no comprehensive consideration for each type of ship, which cannot effectively verify the results. Liu et al. [30] proposed a nonlinear robust control algorithm based on Backstepping method to control ship straight-path tracking. However, there is no curve path tracking method in this study. Zhao et al. [31] based on Serret-Frenet frame transformation develop a tracking error model and the backstepping controller compensates the nonlinearity of the container ship. Xu et al. [32] proposed an adaptive backstepping controller for path-following control of an underactuated ship based on a nonlinear steering model, which can achieve the effect of path control. But the shortcoming in both two is, they present pretty few control scenarios of the researches to verify the reliability of the algorithm, moreover, the scene of track control is too simple.

This paper presents a new path-following control law. The main innovation of this paper is to propose hyperbolic guidance law and apply it to straight-line path-following control. This design improves the convergence of the straight-line path-following control. For curved path-following in the transition between two adjacent straight-line paths, this paper improved a reverse stepping method to calculate heading rate commands to make the system globally asymptotically stable. Compared with the previous research, the extended state observer (ESO) for the controller model is established to estimate and compensate the state in this paper, which can improve the anti-interference ability of the ship in the path following. In the presence of sea current interference, also gives a correction formula. In this paper, linear quadratic regulator (LQR) and nonlinear feedback control are combined to improve the traditional LQR heading controller to improve the control accuracy of heading and yaw rate of the ship. Finally, the track control effect of three different types of ships is verified in simulations and the control effect is evaluated according to International Standard IEC62065 [33].

Section 2 establishes a mathematical model, including ship motion simulation and control and identifies controller parameters. Section 3 describes the design of the heading controller. Section 4 introduces the new guidance algorithm. Section 5 presents analysis of simulation results. Finally, Section 6 contains the conclusions.

2. Mathematical Model of Ship Motion

2.1. Process Plant Model

We describe the ship motion model in five basic compositions, with three degrees of freedom, shown in Figure 1. The model is derived from Newton's laws of motion using linear equations to relate hydrodynamic forces to the respective motions in the horizontal plane. The equations are as follows [4]:

$$m\left[\dot{u} - vr - x_g r^2 - y_g \dot{r}\right] = X, \tag{1}$$

$$m\left[\dot{v} + ur - y_g r^2 + x_g \dot{r}\right] = Y, \tag{2}$$

$$I_z \dot{r} + m\left[x_g(\dot{v} + ur) - y_g(\dot{u} - vr)\right] = N, \tag{3}$$

where X, Y and N denote external forces and moment, while (x_g, y_g, z_g) describe the location of the center of gravity. u,v and r denote surge velocity, sway velocity and yaw angular velocity, respectively, $\dot{u}$,$\dot{v}$ and $\dot{r}$ are their derivatives. I_z is the yaw moment of inertia, m is the total mass of ship. It is common to let the body frame coordinate origin coincide with the center of gravity; thus, the equations can be rewritten as:

$$m(\dot{u} - vr) = X, \tag{4}$$

$$m(\dot{v} + ur) = Y, \tag{5}$$

$$I_z \dot{r} = N. \tag{6}$$

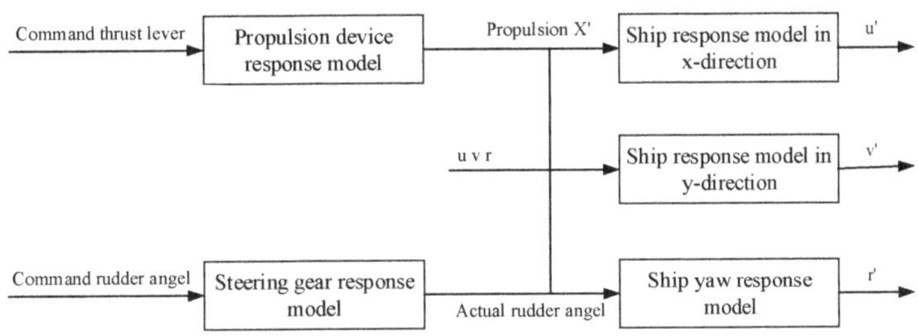

Figure 1. Ship motion mathematical model.

The hydrodynamic force X is linearized and is assumed to be proportional to the ship linear velocity, u in surge. Let R_u denote the linear coefficient of hydrodynamic resistance to forward motion and X_{thrust} denote the thrust provided by the propeller. Thus Equation (4) becomes:

$$m(\dot{u} - vr) = X_{thrust} - R_u u. \tag{7}$$

Let τ_u denote the time constant of the linear surge response model, K_u denote the coefficient of thrust, $u_{\max}$ denote maximum forward speed, and $X_{\max}$ denote maximum thrust:

$$\begin{aligned} \tau_u &= m/R_u, \\ u_{\max} &= X_{\max}/R_u, \\ K_u &= u_{\max}/\tau_u. \end{aligned} \tag{8}$$

Substituting Equation (8) into Equation (7) yields:

$$\dot{u} = K_u X' + vr - u/\tau_u. \tag{9}$$

In the same way, for sway response model let $\tau_v = m/R_v$ denote the linear coefficient of hydrodynamic resistance to sway motion, then Equation (5) becomes:

$$\dot{v} = -ur - v/\tau_v. \tag{10}$$

For the yaw response model, Equation (6) becomes:

$$I_z\dot{r} = K_r(\delta_a K_u X'\tau_u/L + W) + \gamma L R_v(v - \gamma L r) - R_r r, \quad (11)$$

where K_r is a constant of proportionality, γ is the yaw stability factor, L is the length of the ship, δ_a is the real rudder deflection, R_v is the linear coefficient of hydrodynamic resistance to lateral motion, R_r is the linear coefficient of hydrodynamic resistance to yaw and $\tau_r = I_z/R_r$ denotes the time constant of the yaw response model. Simplifying by relating the moment of inertia to the mass and length as a uniformly dense rod to its mass and length, then the moment of inertia can be obtained as $I_z = mL^2/12$. $K_r W$ is the turning moment imparted by wave disturbance, K'_r is the normalized coefficient of rudder moment, which can be expressed as $K'_r = K_r\delta_{\max}/I_z$.

The wave interference model established in this paper consists of a series of square half wave waves. The duration T and height H of each half wave are random numbers related to the Bretschneider wave spectrum [34]. Finally, a square wave interference signal is generated by the wave model, in which the calculation methods of the duration T and the wave height H are as follows:

$$\begin{cases} T = 0.5T_0(ss) \times (1 + 0.5Rand_b) \\ H = H_0(ss) \times (1 + H_r \times Rand_b) \times A_a \end{cases}, \quad (12)$$

where $Rand_b$ denotes a random number among $[-1, 1]$. A value of 0.5 for H_r is recommended, while A_a has alternating sign (i.e., ± 1) to describe wave direction, ss means the sea state level, $T_0(ss)$ and $H_0(ss)$ are the natural period and significant wave height under sea state ss, respectively. The range of values is shown in Table 1.

Table 1. The value of natural period and meaningful wave height under different sea conditions.

Sea State (ss)	T_0 (ss)	H_0 (ss)
1	2.2	0.1
2	5	0.5
3	7.8	1.25
4	11	2.5
5	14	4
6	17.2	6
7	21.1	9
8	26.3	14

Referring to the modeling method of wave interference in IEC62065, this paper ignores the influence of wave interference on ship surge motion and sway motion, only adds the wave interference torque of ship's yaw motion to the mathematical model of ship's motion. Rewrite ship yaw Equation (11) as:

$$\dot{r} = K'_r \frac{K_u X'\tau_u}{L} \cdot \delta_a' + \frac{12\gamma(v - \gamma L r)\tau_r}{L\tau_v} - \frac{r}{\tau_r} + 0.01K'_r S_f H, \quad (13)$$

where δ_a' is the normalized rudder position obtained by $\delta'_a = \delta_a/\delta_{a\max}$, S_f is a Scaling Factor. A value of 20 for the Scaling Factor is recommended.

2.2. Control Plant Model

We simplify the process plant model described in the previous section to design the autopilot control algorithm. The simplified model contains only the main physical properties of the process. Notomo [35] proposed a linearized model for ship steering equations, which is given by a simple transfer function between r and δ.

$$\frac{r}{\delta}(s) = \frac{K}{(1 + Ts)}, \quad (14)$$

where T and K are the time constant and gain constant, respectively. Neglecting the roll and pitch modes ($\phi = \theta = 0$), such that:

$$\dot{\psi} = r \tag{15}$$

finally yields:

$$\frac{\psi}{\delta}(s) = \frac{K}{s(1+Ts)}, \tag{16}$$

where ψ is the heading angle.

The time domain form of Equation (16) is:

$$T\ddot{\psi} + \dot{\psi} = K\delta. \tag{17}$$

Notomo's first order model is usually written as:

$$T\dot{r} + r = K\delta. \tag{18}$$

When the straight motion of the ship is unstable or critical stable, nonlinear maneuvering models should be used. A nonlinear term can be added in Equation (18); thus, Notomo's nonlinear first order model can be obtained as:

$$T\dot{r} + r = K(\delta + K_v v). \tag{19}$$

This is the commonly used control plant model for the design of the steering autopilot, in which K_v is the coefficient of the nonlinear term.

2.3. Controller Parameter Identification

In the above-mentioned control model, some parameters cannot be obtained directly but need to be obtained by system parameter identification. In control applications, the controller performance is bound to the assumptions and approximations of the model. To obtain the corresponding controller parameters more accurately, we apply the forgetting factor least square algorithm to Notomo's nonlinear first order model to identify the ship motion model.

The Recursive Least Square (RLS) algorithm with exponential forgetting is given as follows [36]:

$$\hat{\theta}(k) = \hat{\theta}(k-1) + Q(k)\left[y(k) - \Phi^T(k)\hat{\theta}(k-1)\right], \tag{20}$$

$$Q(k) = P(k-1)\Phi(k)\left[\lambda I + \Phi^T(k)P(k-1)\Phi(k)\right]^{-1}, \tag{21}$$

$$P(k) = \frac{1}{\lambda}\left[I - Q(k)\Phi^T(k)\right]P(k-1), \tag{22}$$

where the forgetting factor value of λ is $0.98 < \lambda < 0.995$, $\Phi(k)$ is the data vector and $\hat{\theta}(k)$ is the estimated value of the parameter vector. $Q(k)$ and $P(k)$ are the intermediate variable help us get $\hat{\theta}(k)$ we want.

Next, referring to the ship motion simulation model, substituting control Equation (19) into ship motion Equation (13), the values of T, K and K_v are obtained as follows:

$$\begin{cases} T = \frac{\tau_v \tau_r}{\tau_v + 12\tau_r\gamma^2}, \\ K = \frac{TK_r' u_{\max} X'}{L\delta_{\max}}, \\ K_v = \frac{12\gamma T}{KL\tau_v}. \end{cases} \tag{23}$$

To facilitate the identification calculation, Equations (11) and (13) are rewritten in the following form:

$$\begin{cases} Y_0 = \theta_0 X_0 + \theta_1 X_1 \\ Y_1 = \theta_2 X_2 \end{cases} \tag{24}$$

where

$$\begin{cases} Y_0 = \dot{r} - \frac{12\gamma}{L\tau_v}v_w + \frac{12\gamma^2}{\tau_v}r \\ Y_1 = \dot{v}_g + u_g r \end{cases}, \tag{25}$$

$$\begin{cases} X_0 = \frac{U}{L}\delta'_a \\ X_1 = r \\ X_2 = v_w \end{cases}, \tag{26}$$

$$\begin{cases} \theta_0 = \frac{KL}{TU} \\ \theta_1 = -\frac{1}{T} + \frac{12\gamma^2}{\tau_v} \\ \theta_2 = -\frac{1}{\tau_v} \end{cases}. \tag{27}$$

According to the recursive formula of the forgetting factor least squares identification algorithm in Equations (20)–(22), the corresponding identification results θ_0, θ_1 and θ_2 can be obtained. Through θ_0, θ_1 and θ_2, the three controller parameters $\hat{T}$, $\hat{K}$ and $\hat{K}_v$ of the nonlinear first-order Notomo model can be solved:

$$\begin{cases} \hat{T} = \frac{1}{-\hat{\theta}_1 + \frac{12\gamma^2}{\tau_v}}, \\ \hat{K} = \frac{\hat{\theta}_0 \hat{T} U}{L}, \\ \hat{K}_v = \frac{12\gamma\hat{T}}{L\hat{K}\tau_v}. \end{cases} \tag{28}$$

3. Optimal Heading Controller

This section describes the design of the optimal heading controller. The main goal is to calculate the controller parameters in terms of the Notomo constants obtained by the identification algorithm in the previous section and introduce them in the LQR controller law. Then, nonlinear feedback is added to improve LQR controller to achieve nonlinear control and better control accuracy. In addition, to filter out the influence of sea wave interference on the acquisition of ship heading signal, an extended state observer (ESO) is added to this design to obtain the estimated value of the low-frequency heading signal. Thus, the controller can obtain accurate and fast course control.

3.1. LQR Controller

Assume that the heading signal ψ can be obtained by a compass and the yaw rate r can be obtained by a rate gyro or by a state observer. Notomo's first order model in Equation (18) can be written as the state space form:

$$\dot{X} = AX + Bu_c, \tag{29}$$

where $X = (r, \psi, \delta)^T$ is the state vector and u_c is the controller input, which denotes the desired rudder rate. A and B are the coefficient matrixes with:

$$A = \begin{pmatrix} -1/T & 0 & K/T \\ 1 & 0 & 0 \\ 0 & 0 & 0 \end{pmatrix}, \tag{30}$$

$$B = \begin{pmatrix} 0 & 0 & 1 \end{pmatrix}^T, \tag{31}$$

$X_d = (r_d, \psi_d, \delta)^T$ denotes the objective state vector. The goal of the controller design is to converge the state X to the objective X_d:

$$\lim_{t\to\infty} X = X_d, \tag{32}$$

$\widetilde{X}$ denotes the error vector:

$$\widetilde{X} = X - X_d = (\widetilde{r}\,\widetilde{\psi}\,\delta)^T, \tag{33}$$

where $\tilde{r} = r - r_d$, $\tilde{\psi} = \psi - \psi_d$. The control performance specification can be measured in terms of:

$$J = \int_0^{\infty} \left(\tilde{X}^T Q X + u_c{}^T R u_c \right) dt, \quad (34)$$

where Q is a 3-dimensional diagonal matrix, in which diagonal elements are the weighting factors. R is the weighting factor of the input. The LQR problem is to find the optimal control $u_c(t)$ such that J in Equation (34) is minimized,

$$u_c = -R^{-1} B^T P \tilde{X}, \quad (35)$$

where P is the positive defined solution of the Riccati Equation [37]:

$$A^T P + PA - PBR^{-1} B^T P + Q = 0 \quad (36)$$

Let

$$(k_1, k_2, k_3)^T = -R^{-1} B^T P, \quad (37)$$

where k_1, k_2 and k_3 are the controller gains obtained by the LQR law. Thus, the representation of the controller can be written as:

$$\delta_{LQR} = \int_0^t (k_1 \tilde{r} + k_2 \tilde{\psi} + k_3 \delta) d\tau. \quad (38)$$

3.2. Feedback Nonlinearization Compensation

In the controller design here, a nonlinear feedback term is added in the control law for compensating the nonlinear maneuvering of the ship. Notomo's nonlinear first order model can be rewritten as:

$$T\dot{r} + r = K(\delta + K_v v). \quad (39)$$

Comparing with the linear controller law obtained in the previous section, the following equation can be obtained:

$$\delta_{LQR} = \delta_{com} + K_v v, \quad (40)$$

where δ_{com} is the command rudder angle, $K_v v$ can be treated as the feedback term, such that:

$$\delta_{com} = \delta_{LQR} - \delta_{FL}, \quad (41)$$

with $\delta_{FL} = K_v v$ and the value of K_v is obtained from the identification algorithm in the previous section.

3.3. Extended State Observer

Both surface ships and underwater vehicles need state observers to process signal data from sensors and navigation equipment. The observer in this paper is designed in terms of the nonlinear ship model and the wave disturbance model. The nonlinear ship model and the wave disturbance model are as follows:

$$\begin{cases} T\dot{r} + r = K(\delta - \delta_n + K_v v) + \omega_r \\ \ddot{\psi}_\omega + 2\varsigma\omega_0 \dot{\psi}_\omega + \omega_0^2 \psi_\omega = \omega_\omega \end{cases} \quad (42)$$

where δ_n is the rudder offset, ω_r and ω_ω are the zero-mean Gaussian measurement white noise; ψ_ω represents the first-order wave-induced motion, and ς is the relative damping ratio, conventionally assigned a value of 0.075. ω_0 is the wave frequency.

Based on Equation (42), the state space equation of the extended state observer is:

$$\dot{\hat{X}} = A\hat{X} + B\Gamma + G(Y - C\hat{X}), \quad (43)$$

where $\hat{X}$ denotes the estimation of the state, $\hat{X} = (\hat{\delta}_n, \hat{r}, \psi, \hat{\xi}_\omega, \hat{\psi}_\omega)^T$ and $\dot{\hat{\xi}}_\omega = \hat{\psi}_\omega$. Γ is the input of the model and $\Gamma = \delta + K_v v$; G is the gain matrix; $\hat{\delta}_n$ is the estimation of the rudder offset; Y is the heading signal obtained by the compass with $Y = \psi + \psi_\omega$.A, B and C are the coefficient matrixes with:

$$A = \begin{pmatrix} 0 & 0 & 0 & 0 & 0 \\ -\frac{K}{T} & -\frac{1}{T} & 0 & 0 & 0 \\ 0 & 1 & 0 & 0 & 0 \\ 0 & 0 & 0 & -\omega_0^2 & -2\varsigma\omega_0 \end{pmatrix}, \tag{44}$$

$$B = \begin{pmatrix} 0 \\ \frac{K}{T} \\ 0 \\ 0 \\ 0 \end{pmatrix}, \tag{45}$$

$$C = \begin{pmatrix} 0 \\ 0 \\ 1 \\ 0 \\ 1 \end{pmatrix}. \tag{46}$$

The error of the state vector can be denoted as $E = \hat{X} - X$, of which the differential equation is:

$$\dot{E} = (A - GC)E. \tag{47}$$

To keep the asymptotic stability of the error vector, the solution of Equation (47) should be $E(t) = E_0 e^{-kt}$, where E_0 is the initial error value and k is a scale factor. Thus, all of the eigenvalues of the matrix $(A - GC)$ have negative real parts. The gain matrix G can be found by pole placement in terms of the eigenvalues. In this case, the eigenvector of the matrix $(A - GC)$ is chosen as: $P = (P_0, P_1, P_2, P_3, P_4) = (1.5/T, 1.5/T, 1.5/T, 15\varsigma\omega_0, 15\varsigma\omega_0)$ and the gain matrix can be obtained by using the MATLAB function.

4. Guidance Law

Systems for guidance consist of a waypoint generator with human interface. A new hyperbolic guidance method for straight-line path-following control is presented in this paper. A Lyapunov function analysis [38] is used to prove the stability of this method. For curve guidance, based on the Lyapunov stability function, this paper presents a reverse stepping method to calculate heading rate command and make the system globally asymptotically stable, with a correction formula for current interference.

4.1. Error Coordinates

Consider a straight-line path implicitly defined by two waypoints $p_k = (x_k, y_k)^T \in R^2$ and $p_{k+1} = (x_{k+1}, y_{k+1})^T \in R^2$, respectively. Also, consider the position of the ship denoted by the point $p_t = (x_t, y_t)^T \in R^2$. Then the direction of the path can be defined as:

$$\alpha_k := \text{atan2}(x_{k+1} - x_k, y_{k+1} - y_k). \tag{48}$$

Fossen [4] gives the formula to calculate the cross-track error:

$$\widetilde{P}(t) = -[x_t - x_k]\cos(\alpha_k) + [y_t - y_k] sin(\alpha_k). \tag{49}$$

The calculation of the cross-track error in the transition curved path between two adjacent straight-line paths is complicated. Figure 2 shows two adjacent straight-line paths described by the waypoints p_k,p_{k+1} and p_{k+2}. R_{k+1} is the radius of the transition curve and the cross-track error of the transition curved path can be obtained by:

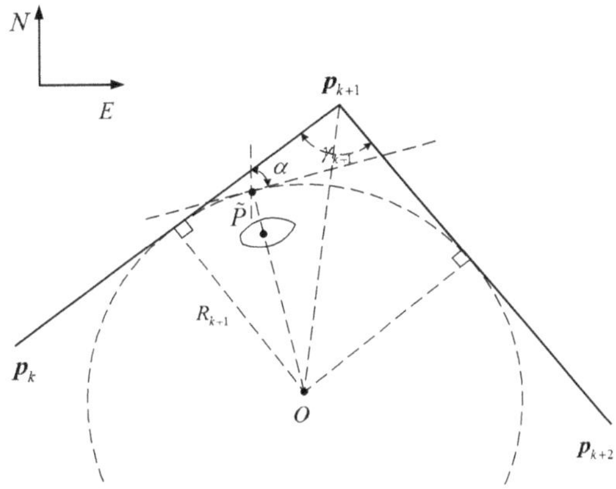

Figure 2. The transition curved path.

First, determine the direction of the curved path (clockwise or counterclockwise). Based on Equation (48), the direction α_k of the straight-line path $p_k p_{k+1}$ and the direction α_{k+1} of the straight-line path $p_{k+1}p_{k+2}$ can be calculated. Then the direction of the curved path is:

$$F_{k+1} = \text{sign}(sin(\alpha_{k+1} - \alpha_k)); \tag{50}$$

here, F_{k+1} denotes the direction of the curved path, a positive sign means clockwise and a negative sign means counterclockwise.

Second, determine the center point of the curved path. The angle between the two adjacent straight-line paths can be obtained in terms of the direction α_k, α_{k+1} and F_{k+1}.

$$\gamma_{k+1} = \pi - F_{k+1}(\alpha_{k+1} - \alpha_k). \tag{51}$$

Then the distance between the center point and the waypoint p_{k+1} can be determined in terms of the angle γ_{k+1} and the radius R_{k+1}.

$$D_{k+1} = R_{k+1}/\sin\left(\frac{\gamma_{k+1}}{2}\right). \tag{52}$$

The angle between the vector from the center point O_{k+1} to the waypoint p_{k+1} and North can be obtained in terms of the angle γ_{k+1}, the direction F_{k+1} and α_k.

$$\alpha_{O_{k+1}} = \alpha_k - F_{k+1}\gamma_{k+1}. \tag{53}$$

Based on the waypoint p_{k+1}, the direction $\alpha_{O_{k+1}}$ and the distance D_{k+1} from the center point O_{k+1} to the waypoint p_{k+1}, the coordinate $P_{O_{k+1}} = \left(x_{O_{k+1}}, y_{O_{k+1}}\right)$ of the center point O_{k+1} can be determined:

$$\boldsymbol{p}_{O_{k+1}} = \boldsymbol{p}_{k+1} - R_{k+1}\begin{pmatrix} \sin\alpha_{O_{k+1}} \\ \cos\alpha_{O_{k+1}} \end{pmatrix}. \tag{54}$$

Finally, the cross-track error can be calculated by:

$$\widetilde{P}(t) = F_{k+1}\|\boldsymbol{p}_t - \boldsymbol{p}_{O_{k+1}}\| - R_{k+1}. \tag{55}$$

The course angle command is the direction of the vector tangential to the point on the path that is closest to the vehicle. The course angle command can be calculated by:

$$\widetilde{\chi}(t) = \chi(t) - atan2\left(x(t) - x_{O_{k+1}}, y(t) - y_{O_{k+1}}\right) + \frac{\pi}{2}F_{k+1}; \tag{56}$$

here, $\chi(t)$ is the current course.

4.2. Straight-Line Path Guidance

The main goal of straight-line path-following control is to eliminate cross-track and heading errors. Hyperbolic guidance will achieve a more rapid convergence of path-following maneuvering because of its smoothing and transition properties. Thus, a hyperbolic guidance methodology is adopted with a hyperbolic tangent function. As shown in Figure 3, the curve OC denotes a hyperbola, of which the asymptote is the straight-line path. The hyperbola equation is:

$$y = R_c \frac{e^{k_c x} - e^{-k_c x}}{e^{k_c x} + e^{-k_c x}}, \tag{57}$$

where R_c denotes the distance between the origin O and the straight-line and $\widetilde{P}(t)$ is the distance from the ship to the planned route at time t with $R_c > \widetilde{P}(t)$.

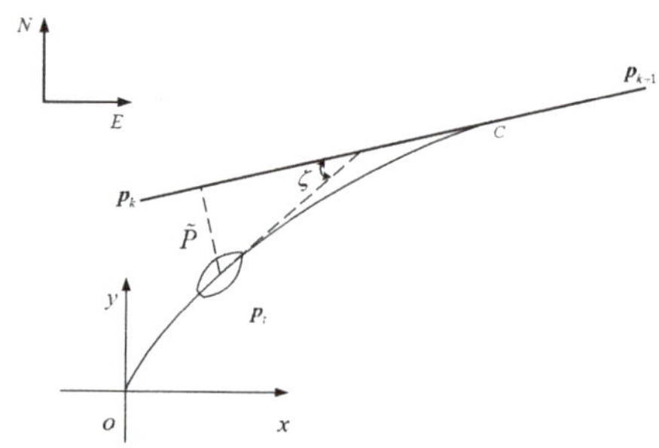

Figure 3. The hyperbolic guidance law.

Then, the position of the ship p_t can be denoted by $\left(x_0, R_c - \widetilde{P}(t)\right)$ and the angle ζ shown in Figure 3 can be calculated by:

$$\zeta(t) = \text{sign}\left(\widetilde{P}(t)\right)\arctan\left[k_c\left(2\left|\widetilde{P}(t)\right| - \frac{\widetilde{P}(t)^2}{R_c}\right)\right], \tag{58}$$

where k_c is a constant coefficient with $k_c > 0$, denoting the rate of approach to the straight-line path.

Then the course command can be obtained by:

$$\chi_c = \alpha_k + \zeta(t) = \alpha_k + \text{sign}\left(\widetilde{P}(t)\right)\arctan\left[k_c\left(2\left|\widetilde{P}(t)\right| - \frac{\widetilde{P}(t)^2}{R_c}\right)\right], \tag{59}$$

where χ_c is the command course and α_k is initial heading angle.

4.3. Curved Path Guidance

For curved path following, a kinematic controller generates the desired states for motion control. The control method can be designed using a dynamic model of the ship by specifying a reference frame that moves along the path. A Serret-Frenet reference frame [39] is usually chosen. Figure 4 shows an inertia reference frame $\{i\} = (x_i, y_i, z_i)$, a body-fixed reference frame $\{s\} = (x_s, y_s, z_s)$ and a Serret-Frenet frame $\{m\} = (x_m, y_m, z_m)$. The origin O_m of $\{m\}$ is attached to the point on the path that is closest to the vehicle. R denotes the radius of the curved path.

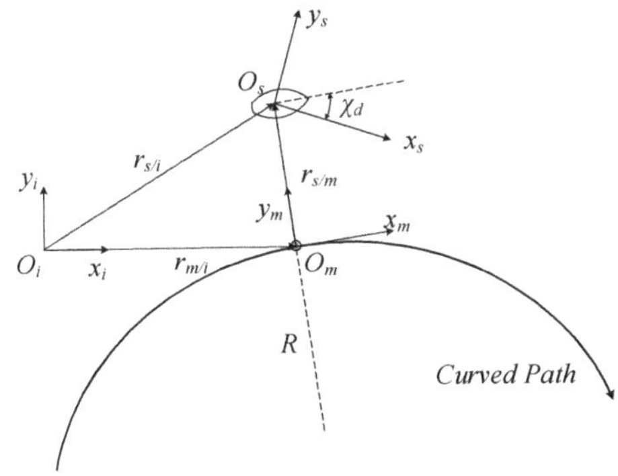

Figure 4. Path-following: reference frame.

The notation $H^c_{a/b}$ is adopted in this paper, where $\{a\}$, $\{b\}$ and $\{c\}$ denote the three different frames and H denotes the coordinate of $\{a\}$ in frame $\{b\}$ relative to $\{c\}$. Such that:

$$r^i_{m/i} + r^i_{s/m} = r^i_{s/i}, \tag{60}$$

where r denotes the distance vector.

Time differentiation of Equation (60) yields:

$$v^i_{m/i} + \frac{d}{dt} r^i_{s/m} = v^i_{s/i}, \tag{61}$$

where v denotes the speed vector. And $\frac{d}{dt} r^i_{s/m}$ can be obtained by:

$$\frac{d}{dt} r^i_{s/m} = R^i_m \omega^i_{m/i} \times r^m_{s/m} + R^i_m v^m_{s/m}, \tag{62}$$

where R^i_m denotes the transformation matrix from $\{m\}$ to $\{i\}$ and $\omega^i_{m/i} = (0, 0, -\dot{\alpha})^T$ denotes the rotation vector between the $\{m\}$ and $\{i\}$ frames.

Substituting Equation (62) into Equation (61) and multiplying by R^i_m on both sides of the equation, yields:

$$v^m_{m/i} + \omega^i_{m/i} \times r^m_{s/m} + v^m_{s/m} = R^m_s v^s_{s/i}. \tag{63}$$

Then the kinematic equations can be obtained:

$$\begin{cases} \dot{\widetilde{P}} = -u \sin \widetilde{\chi} \\ \dot{\alpha} = \frac{u \cos \widetilde{\chi}}{R + F\widetilde{P}} \end{cases}, \tag{64}$$

where $\widetilde{P}$ is the position deviation, $\widetilde{\chi}$ is the course deviation, α is the planned course and F indicates the direction of the transition curve. $F = 1$ means turning clockwise, $F = -1$ means turning counterclockwise.

Thus, goal of the curved path-following controller is to calculate the yaw rate commands, forcing the ship to approach the curved path. This paper presents a reverse stepping calculation method based on the Lyapunov function to obtain the yaw rate commands.

First, calculate the course commands. Consider the Lyapunov function candidate as follows:

$$V_1 = \frac{1}{2} \widetilde{P}^2. \tag{65}$$

Time differentiation of Equation (65) yields:

$$\dot{V}_1 = -\widetilde{P} u \sin(\chi - \alpha). \tag{66}$$

To satisfy global asymptotic stability (GAS), it is necessary to guarantee that $V_1 < 0$ as $\widetilde{P} \neq 0$ and $\dot{V}_1 = 0$ as $\widetilde{P} = 0$. Thus, the following equation is adopted:

$$\dot{V}_1 = -k_1\widetilde{P}^2 \leq 0, \tag{67}$$

where k_1 is the scaling factor, with $k_1 > 0$.

Based on Equation (65) and Equation (66), the course commands can be calculated by:

$$\chi_c = \alpha + \arcsin\left(\frac{k_1\widetilde{P}}{u}\right). \tag{68}$$

Equation (68) suggests that the course command χ_c will approach the path direction α to eliminate the course error $\widetilde{\chi}$, as the cross-track error is decreasing. However, the path direction α is varying with ship motion during the curved path-following control. Thus, it is necessary to calculate the yaw rate commands. Consider the Lyapunov function candidate:

$$V_2 = \frac{1}{2}\widetilde{\chi}^2, \tag{69}$$

with $\widetilde{\chi} = \chi - \chi_c$. Time differentiation of Equation (69) yields:

$$\dot{V}_2 = (\chi - \chi_c)\left(\dot{\chi} - \dot{\alpha} - \frac{\dot{\widetilde{P}}}{\sqrt{(u/k_1)^2 - \widetilde{P}^2}}\right). \tag{70}$$

To satisfy GAS, it is necessary to guarantee that $\dot{V}_2 < 0$ as $\chi = \chi_c$ and $\dot{V}_2 = 0$ as $\chi = \chi_c$. Thus, the following equation is adopted:

$$\dot{V}_2 = -k_2\widetilde{\chi}^2, \tag{71}$$

where k_2 is the scaling factor, with $k_2 > 0$.

Based on Equations (64), (68), (70) and (71), the course commands can be calculated by:

$$\dot{\chi}_c = \frac{u\cos\widetilde{\chi}}{R + F\widetilde{P}} - \frac{-u\sin\widetilde{\chi}}{\sqrt{(u/k_1)^2 - \widetilde{P}^2}} - k_2\widetilde{\chi}, \tag{72}$$

where $\dot{\chi}_c$ is the derivative of χ_c, indicating the command track angular rate. We neglect the varying of the drift angle β, thus the varying rate course command can be treated as the yaw rate command, approximately. However, this does not fit in the presence of wave disturbance. In the presence of sea current interference, the yaw rate commands need to be adjusted as:

$$r_c = \dot{\chi}_c\frac{u_g}{u_g - u_f\cos(\varphi_g - \varphi_f)}, \tag{73}$$

where r_c denotes the yaw rate command, u_g denotes the ship speed over the ground, with direction φ_g and u_f is the current speed through water, with direction φ_f.

Figure 5 is the three-layer control structure of ship track control proposed in this paper.

The track control module calculates the current planned navigation section of the ship according to the current ship position information and environmental interference information and calculates the track deviation and heading deviation, then modifies the planned heading angle according to the deviation information, calculates the command heading angle and sends it to the heading control module. According to the current compass signal, rudder angle signal and yaw speed signal, the course control module calculates the course deviation after comparing with the command heading and calculates the command rudder angle through the course control algorithm.

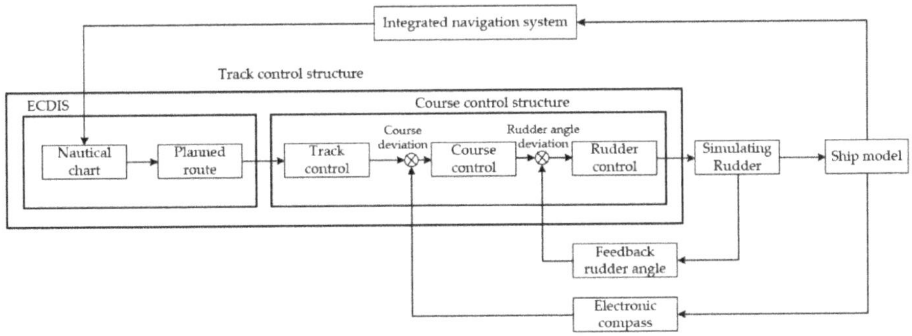

Figure 5. Basic composition of track control system.

Finally, the rudder angle control layer drives the rudder to make the actual rudder angle consistent with the command rudder angle, so as to realize the track control of the ship.

Figure 6 shows the path-following control overall flow chart:

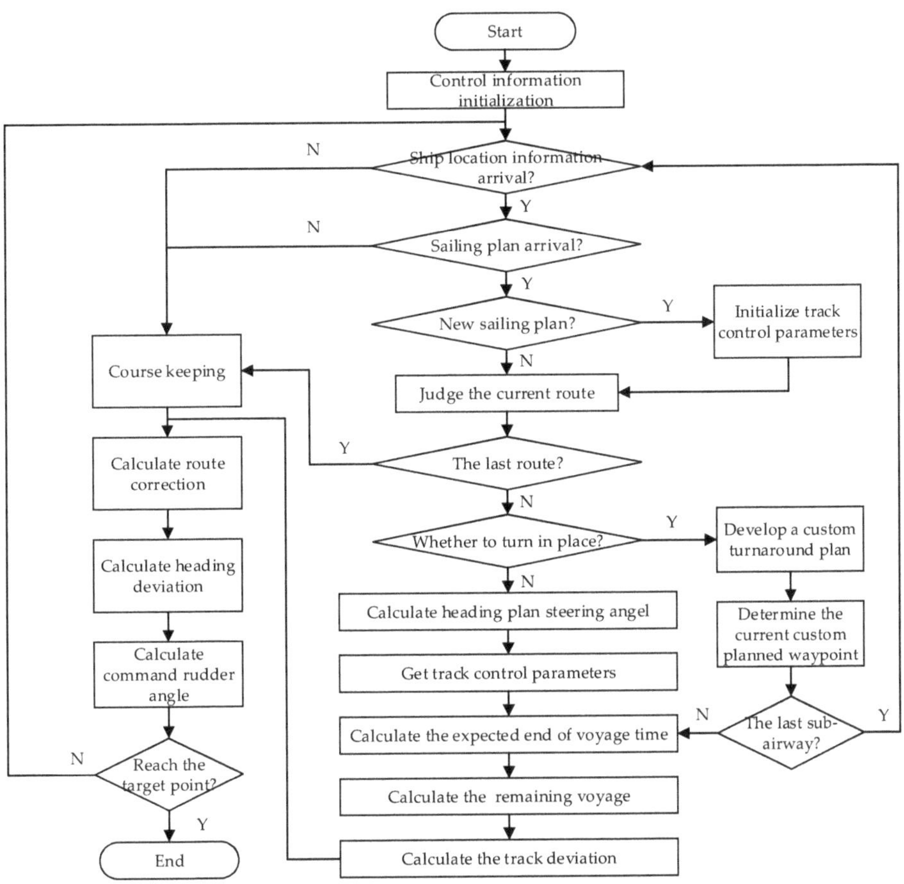

Figure 6. Path-following control flow chart.

5. Simulation Results

International Standard IEC62065 is used to test the performance of the path-following control algorithm presented in this paper. Based on the standard, three classes of ship are adopted: class A fast ferry, class B container and class C tanker. See Table 2 for details.

Table 2. Parameters for three ships.

Parameter	Notation	Ship A Fast Ferry	Ship B Container	Ship C Tanker
Ship length (m)	L	60	250	350
Thrust lever ramp time (s)	T_p	20	30	30
Rudder ramp time (s)	T_δ	12	30	30
Rudder follow-up offset (%)	δ_n	0	0	0
Maximum speed (kn)	u_{max}	30	25	10
Rudder moment coefficient (deg/s/%)	K_r'	0.025	0.01	0.005
Surge response time constant (s)	τ_u	150	600	800
Sway response time constant (s)	τ_v	2	4	36
Yaw response time constant (s)	τ_r	4	23	46
Stability coefficient	γ	−0.05	0	0

The heading controller parameters are given in Table 3:

Table 3. Heading controller parameters.

	$\hat{K}$	$\hat{T}$	$\hat{K}v$	Q	R	k1	k2	k3
Ship A	0.0798	11.1343	1.9899	diag(100, 0.5, 1)	5	3.0261	0.1414	0.2083
Ship B	0.0794	10.8759	1.9863	diag(100, 0.5, 1)	5	2.9936	0.1414	0.2090
Ship C	0.0794	10.8966	1.9913	diag(100, 0.5, 1)	5	2.9964	0.1414	0.2090

The ships' initial states for the three test scenarios are as follows:

Ship A:(u0, v0, r0, ψ0) = (15.4 m/s, 0 m/s, 0 rad/s, 0 deg);

Ship B:(u0, v0, r0, ψ0) = (12.9 m/s, 0 m/s, 0 rad/s, 0 deg);

Ship C:(u0, v0, r0, ψ0) = (5.1 m/s, 0 m/s, 0 rad/s, 65 deg);

In control applications, it is important to validate the controller experimentally, this paper takes ship class B as the simulation object and the ship parameters are shown in Table 2. The step steering signal is selected as the input signal and the rotation experiment is used as the experimental method of controller parameter identification. The trajectory is shown in Figure 7 and the simulation data is shown in Table 4.

Table 4. Simulation data of class B ship rotation experiment.

Data Name	Experimental Data	Actual Data	Error
Rotation diameter (nm)	0.337	0.336	0.001
Fixed length rotation speed (deg/min)	71	71	0
Steady surge speed (kn)	12.9	12.9	0
Steady sway speed (kn)	−1.03	−1.03	0
Tactical cycle diameter (nm)	0.640	0.640	0
Advance (nm)	0.555	0.551	0.004
Departure (nm)	0.325	0.323	0.002

It can be seen from the experimental results that the control model established in this paper meets the ship's turning characteristics and the established model is reliable.

Secondly, the course control algorithms of the three types of ships in Table 2 are verified. For ship class A, the initial speed of the ship is 15.4m/s, the command thrust is 0.67, under sea state 3 and the command heading angle is 30 degrees. As for ship class B, the initial speed is 12.9 m/s, the command thrust is 0.8, under sea state 3 and the command heading angle is 30 degrees. For ship class C, the initial speed of the ship is 5.1m/s, the command thrust is 1, under sea state 3 and the command heading angle is 30 degrees. Figures 8–10 shows that the improved LQR control algorithm in this paper is compared with the traditional LQR algorithm and the simulation experimental chart of course control is obtained. The simulation data is shown in Table 5.

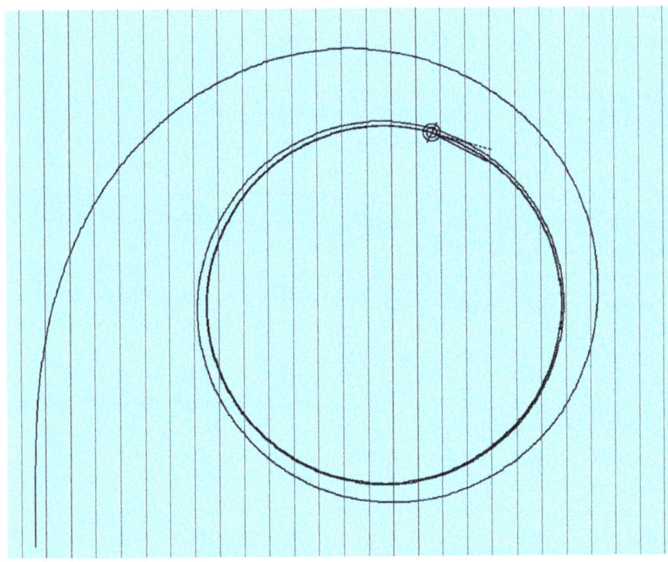

Figure 7. Testing results for ship class B in rotation experiment.

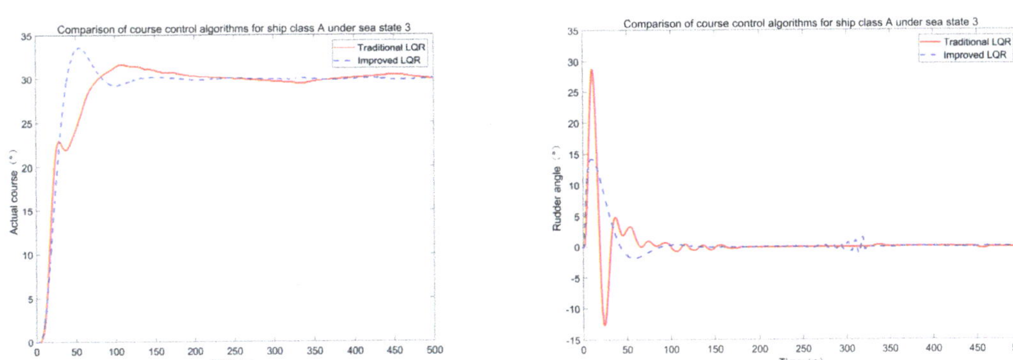

(**a**) (**b**)

Figure 8. Comparative experiment of course control for ship class A: (**a**) The course angle changes with time; (**b**) The actual rudder angle changes with time.

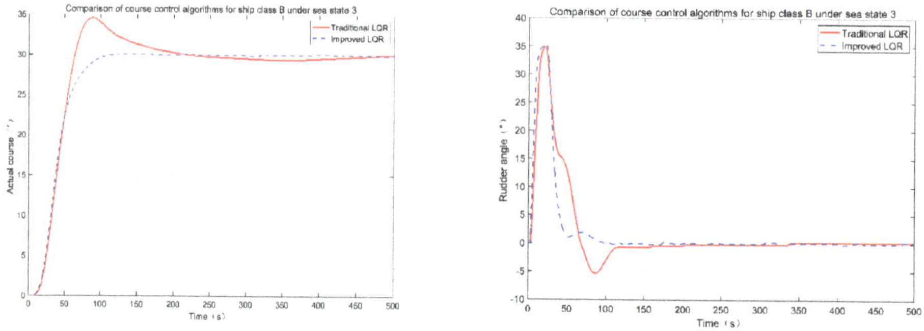

(a) (b)

Figure 9. Comparative experiment of course control for ship class B: (**a**) The course angle changes with time; (**b**) The actual rudder angle changes with time.

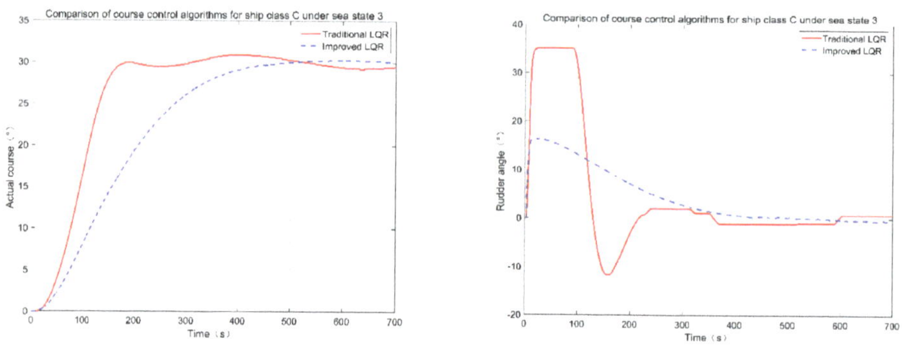

(a) (b)

Figure 10. Comparative experiment of course control for ship class C: (**a**) The course angle changes with time; (**b**) The actual rudder angle changes with time.

Table 5. The results of the heading control test.

Types of Ships	Control Method	Overshoot (°)	Steady State Mean Deviation (°)	Steady State Maximum Deviation (°)	Response Time(s)
Ship class A Fast ferry	Traditional LQR	1.700	0.625	0.500	253.3
	Improved LQR	2.600	0.365	0.200	142.5
Ship class B Container	Traditional LQR	4.600	0.395	0.600	215.1
	Improved LQR	0.100	0.075	0.100	125.7
Ship class C Tanker	Traditional LQR	0.900	0.450	0.900	190.5
	Improved LQR	0.300	0.275	0.100	459.6

Based on the above experimental results, the following conclusions can be drawn: the improved algorithm combining LQR with feedback nonlinearization control can achieve the function of ship course control, the overshoot, control accuracy and corresponding time are better than the traditional LQR control method. In addition, by adding a state observer can ensure that the ship can sail according to the expected instructions under sea state 3 without frequent steering.

There are three test scenarios used to evaluate performance of the path-following controller. Details of the scenarios are given in Tables 6–8.

Table 6. The parameters of the track for ship class A.

Waypoint No.	Latitude	Longitude	Track [deg]	Distance [NM]	Radius [NM]	Estimated ROT [deg/min]
001	00°01.000′ S	000°01.000′ W	000.0	2.00	0.25	80
002	00°01.000′ N	000°01.000′ W	090.0	2.00	0.25	80
003	00°01.000′ N	000°01.000′ E	315.0	1.41	0.10	200
004	00°02.000′ N	000°00.000′ E	225.0	1.41	0.20	100
005	00°01.000′ N	000°01.000′ W	135.0	2.83	0.60	33
006	00°01.000′ S	000°01.000′ E	270.0	2.00	0.20	100
007	00°01.000′ S	000°01.000′ W	045.0	2.83	0.25	80
008	00°01.000′ N	000°01.000′ E	180.0	2.00	0.40	50
009	00°01.000′ S	000°01.000′ E				

Table 7. Parameters of the track for ship class B.

Waypoint No.	Latitude	Longitude	Track [deg]	Distance [NM]	Radius [NM]	Estimated ROT [deg/min]
001	65°00.000′ N	000°20.000′ W	040.2	6.54	0.50	40
002	65°05.000′ N	000°10.000′ W	139.8	13.09	1.0	20
003	64°55.000′ N	000°10.000′ E	040.2	6.55	2.0	10
004	65°00.000′ N	000°20.000′ E				

Table 8. Parameters of the track for ship class C.

Waypoint No.	Latitude	Longitude	Track [deg]	Distance [NM]	Radius [NM]	Estimated ROT [deg/min]
001	00°03.000′ S	179°57.000′ W	000.0	6.00	1.00	10
002	00°03.000′ N	179°57.000′ W	270.0	6.00	1.00	10
003	00°03.000′ N	179°57.000′ E	045.0	4.24	0.50	20
004	00°06.000′ N	180°00.000′ W	135.0	4.24	1.00	10
005	00°03.000′ N	179°57.000′ W	225.0	8.49	1.50	7
006	00°03.000′ S	179°57.000′ E	090.0	6.00	1.00	10
007	00°03.000′ S	179°57.000′ W	315.0	8.49	0.75	13
008	00°03.000′ N	179°57.000′ E	180.0	6.00	1.25	8
009	00°03.000′ S	179°57.000′ E				

Results of testing the three classes of ship are shown in Figures 11–13. WPTs in the figure are waypoints.

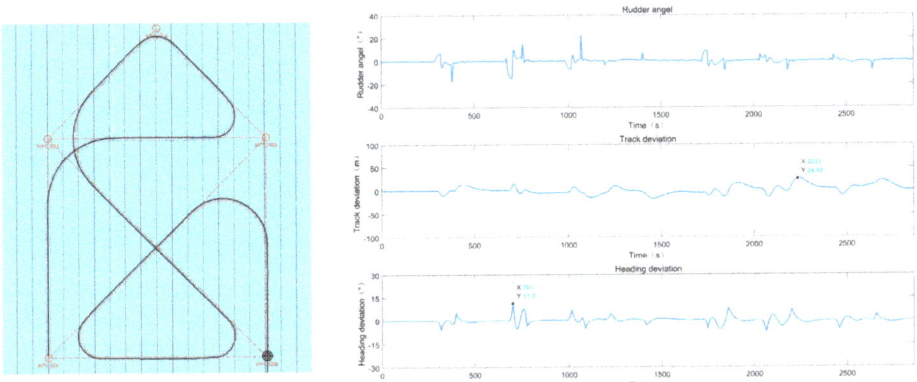

(a) (b)

Figure 11. Testing results for ship class A: (**a**) Ship trajectory; (**b**) Observation index: rudder angle, track deviation and heading deviation.

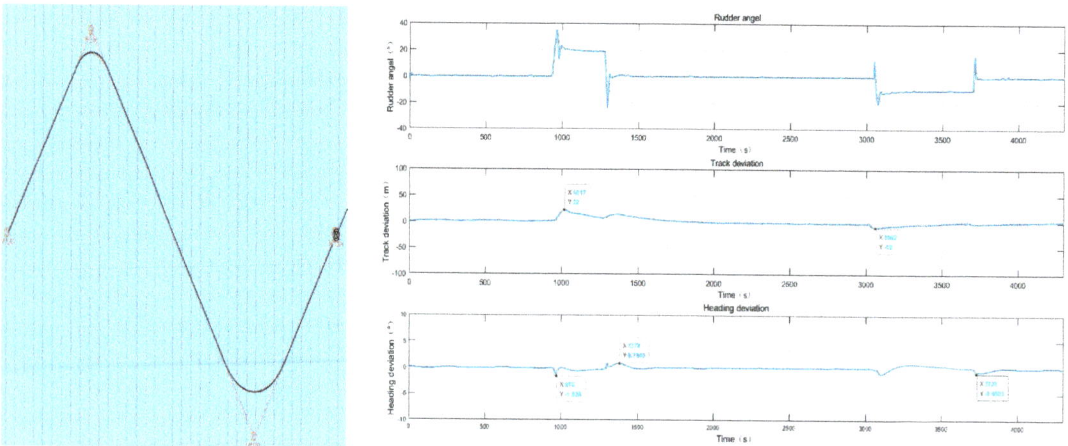

(**a**) (**b**)

Figure 12. Testing results for ship class B: (**a**) Ship trajectory; (**b**) Observation index: rudder angle, track deviation and heading deviation.

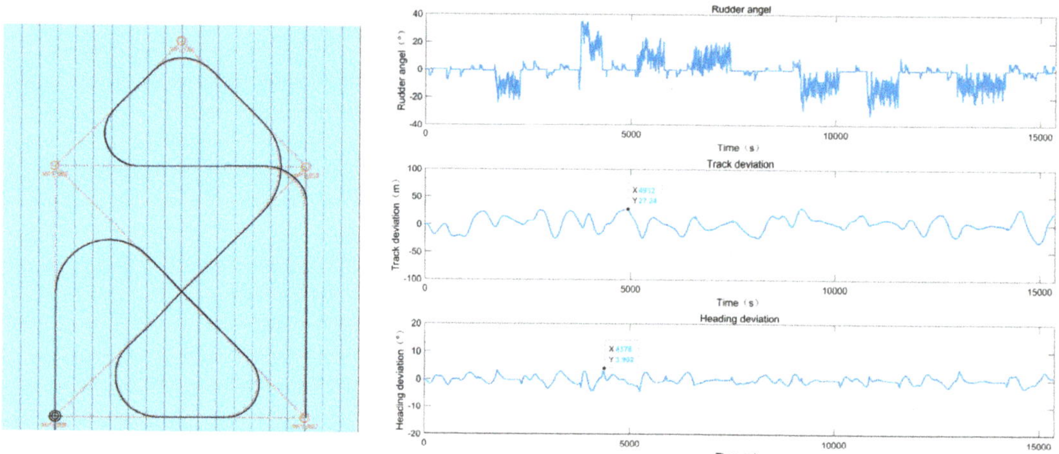

(**a**) (**b**)

Figure 13. Testing results for ship class C: (**a**) Ship trajectory; (**b**) Observation index: rudder angle, track deviation and heading deviation.

From Figure 11, we can see that ship class A has a maximum cross-track error of 24.8 m and the maximum course error 11.3°, which satisfies the requirements set by IEC62065 of setting the course deviation limit to 25° and the cross-track deviation to 100 m. The experimental results show that this algorithm has the advantage of less steering time.

Results for ship class B are shown in Figure 12, with a maximum cross-track error of 22 m and maximum course error of 1.6°, which satisfies the requirements of setting the course deviation limit to 15° and the cross-track deviation to 60 m.

Results for ship class C are shown in Figure 13, where the experimental environment is under rough sea state 5, with maximum cross-track error of 27.24 m and maximum course error of 3.9°, which satisfies the requirements of setting the course deviation limit to 15° and the cross-track deviation to 100 m.

Tests of ship class B in the presence of wave disturbance and a simulated current of 5 kn are also performed. The results are shown in Figure 14.

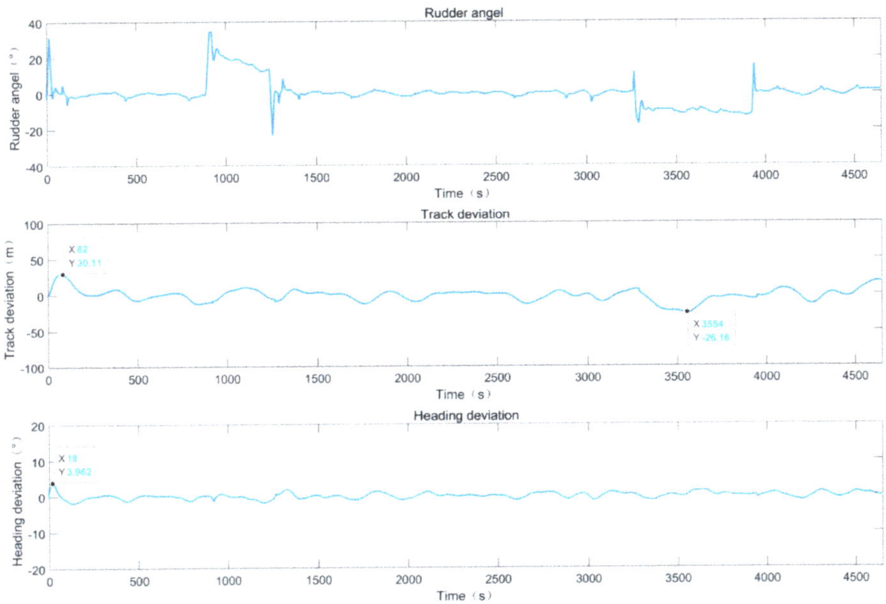

Figure 14. Testing results for ship class B in the presence of disturbance: rudder angle, track deviation and heading deviation.

As shown in Figure 14, in the presence of sea state 3 and a simulated current of 5 kn, 30°N and the other requirements for test are the same as those for ship class B listed earlier, the results for ship class B satisfy the IEC62065 requirements with maximum cross-track error of 30.1 m and maximum course error of 3.96°. The experimental result shows that the proposed algorithm has demonstrated good robustness.

To further verify the control effect of the proposed algorithm, additional comparative experiments have been done. For ship class B, the traditional and adaptive PID control algorithms were used for track control, as shown in Figure 15. The experimental results are also compared with traditional LQR controller and the improved LQR control algorithm with nonlinear feedback term proposed in this paper and the track deviations are shown in Figure 16. Specific data are in Table 9. It can be seen from the experimental results that the traditional LQR control method has better tracking effect in the tracking task of direct sail or low speed tracking simple curve. However, it is difficult to guarantee the tracking accuracy in tracking complex task path. The maximum deviation under traditional PID controller, adaptive PID controller, traditional LQR controller and improved LQR controller are as follows: 176.9 m, 48.51 m, 23.32m and 22 m. In terms of stability, the traditional LQR controller performs poorly. And for the improved LQR controller, the deviation is around 0 for more than 3000 s. The performance of the control method proposed in this paper is significantly better than that of the other three control methods.

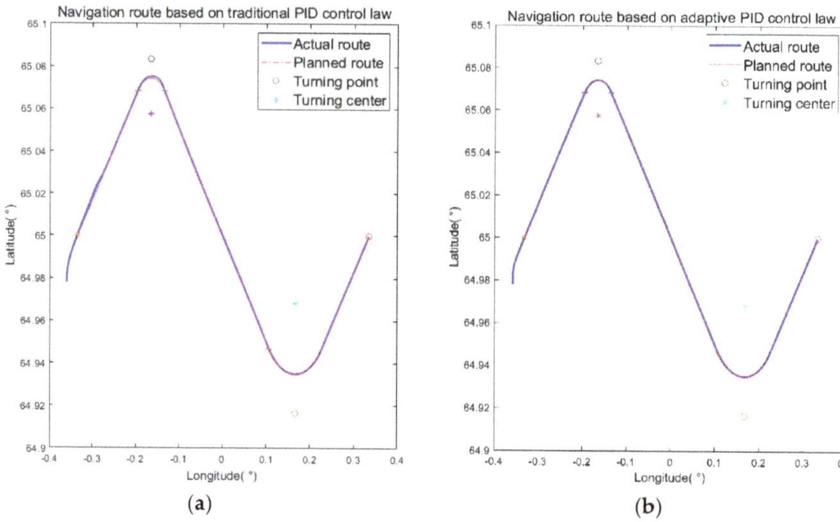

Figure 15. Navigation route for ship class B under different control methods: (**a**) traditional proportional-integral-derivative (PID) control law; (**b**) adaptive PID control law.

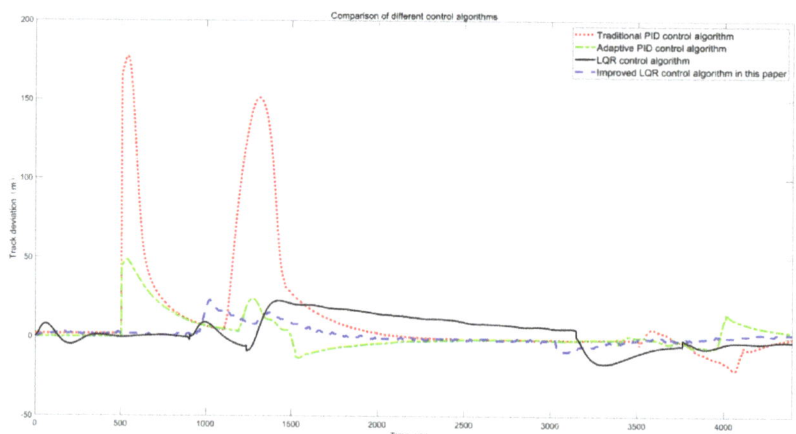

Figure 16. Testing results for ship class B under different control methods.

Table 9. Experimental result for comparison of different control algorithms.

Parameters	Overshoot (m)	Peak Time (s)	Rising Time (s)	Setting Time (s)
Traditional PID	176.9	532, 1305	243	1650
Adaptive PID	48.51	527	135	1080
Traditional LQR	23.32	1406	162	1097
Improved LQR	22	1024	94	690

6. Conclusions

A new practical guidance and control algorithm for marine vehicles is introduced in this paper. The main innovation is the hyperbolic guidance law used in straight-line path-following control. For curved path-following in the transition between two adjacent straight-line paths, this paper improves the reverse stepping method for the controller to calculate heading rate command to make the system globally asymptotically stable, with a correction formula for sea current interference. Traditional LQR and feedback nonlinear control are combined to improve the accuracy of course control. Different types of ships are verified in this paper, simulation results based on the International Standard IEC62065 are included to demonstrate system performance. The algorithms satisfy the IEC62065 testing requirements. Three additional experiments verified that the control method has good robustness and good control effect.

Author Contributions: Conceptualization, Z.Z. and H.W.; methodology, Z.Z.; software, Z.Z.; validation, Z.Z., H.W., Y.Z. (Yuhan Zhao) and G.Z.; formal analysis, Z.Z.; investigation, Z.Z.; resources, Z.Z.; data curation, Y.Z. (Yuhan Zhao) and G.Z.; writing—original draft preparation, Z.Z.; writing—review and editing, Z.Z.; visualization, Z.Z.; supervision, H.W. and Y.Z. (Yi Zhao); project administration, H.W. and Y.Z. (Yi Zhao); funding acquisition, H.W. All authors have read and agreed to the published version of the manuscript.

Funding: This study was funded by the Russian Foundation for Basic Research (RFBR) (No. 20-07-00531).

Institutional Review Board Statement: "Not applicable" for studies not involving humans or animals.

Informed Consent Statement: "Not applicable" for studies not involving humans.

Data Availability Statement: Data sharing not applicable. No new data were created or analyzed in this study. Data sharing is not applicable to this article.

Acknowledgments: The authors are grateful to the anonymous reviewers for their valuable comments and suggestions that helped improve the quality of this manuscript.

Conflicts of Interest: The authors declare no conflict of interest.

References

1. Breivik, M.; Fossen, T.I. Path following for marine surface vessels in Oceans. In Proceedings of the MTS/IEEE Techno-Ocean conference, Kobe, Japan, 9–12 November 2004; pp. 2282–2289.
2. Dong, Z.P.; Wan, L.; Li, Y.M.; Zhang, G.C. Trajectory tracking control of underactuated USV based on modified backstepping approach. *Int. J. Nav. Archit. Ocean Eng.* **2015**, *7*, 817–832. [CrossRef]
3. Filipe, M.B.; Lucas, B.M. Robust path-following control for articulated heavy-duty vehicles. *Control Eng. Pract.* **2019**, *85*, 246–256.
4. Fossen, T.I. *Guidance and Control of Ocean Vehicles*, 1st ed.; Wiley: New York, NY, USA, 1994; pp. 55–122.
5. Breivik, M.; Fossen, T.I. Applying missile guidance concepts to motion control of marine craft. *IFAC Proc. Vol.* **2007**, *40*, 349–354. [CrossRef]
6. Fossen, T.I.; Pettersen, K.Y.; Galeazzi, R. Line-of-sight path following for dubins paths with adaptive sideslip compensation of drift forces. *Control Syst. Technol. IEEE* **2015**, *23*, 820–827. [CrossRef]
7. Kelasidi, E.; Liljeback, P.; Pettersen, K.Y.; Gravdahl, J.T. Integral line-of-sight guidance for path following control of underwater snake robots: Theory and experiments. *IEEE Trans. Robot.* **2017**, *33*, 610–628. [CrossRef]
8. Moe, S.; Pettersen, K.Y. Set-Based line-of-sight (LOS) path following with collision avoidance for underactuated unmanned surface vessels under the influence of ocean currents. In Proceedings of the IEEE Conference on Control Technology and Applications (CCTA), Mauna Lani, HI, USA, 27–30 August 2017; pp. 241–248.
9. Moe, S.; Pettersen, K.Y.; Fossen, T.I.; Gravdahl, J.T. Line-of-sight curved path following for underactuated USVs and AUVs in the horizontal plane under the influence of ocean currents. In Proceedings of the 24th Mediterranean Conference on Control and Automation (MED), Athens, Greece, 21–24 June 2016; pp. 38–45.
10. Fossen, T.I.; Lekkas, A.M. Direct and indirect adaptive integral line-of-sight path following controllers for marine craft exposed to ocean currents. *Int. J. Adapt. Control Signal Process.* **2017**, *31*, 445–463. [CrossRef]
11. Korkmaz, F.C.; Su, M.E.; Alarcin, F. Control of a ship shaft torsional vibration via modified PID controller. *Brodogradnja* **2014**, *65*, 17–27.
12. Samanta, B.; Nataraj, C. Design of intelligent ship autopilots using particle swarm optimization. In Proceedings of the IEEE Swarm Intelligence Symposium, St. Louis, MO, USA, 21–23 September 2008; pp. 1–7.
13. Johansen, T.A.; Fuglseth, T.P.; Tondel, P.; Fossen, T.I. Optimal constrained control allocation in marine surface vessels with rudders. *Control Eng. Pract.* **2008**, *16*, 457–464. [CrossRef]
14. Lee, S.D.; Tzeng, C.Y.; Huang, W.W. Ship streeing autopilot based on ANFIS framework and conditional turing scheme. *Mar. Eng. Front.* **2013**, *1*, 53–62.
15. Dlabac, T.; Calasan, M.; Krcum, M.; Marvucic, N. PSO-Based PID controller design for ship course-keeping autopilot. *Brodogradnja/Shipbuilding* **2019**, *70*, 1–15. [CrossRef]
16. Oh, S.R.; Sun, J. Path following of underactuated marine surface vessels using line-of-sight based model predictive control. *Ocean Eng.* **2010**, *37*, 289–295. [CrossRef]

17. Guerrero, J.; Torres, J.; Creuze, A. Trajectory tracking for autonomous underwater vehicle: An adaptive approach. *Ocean Eng.* **2019**, *172*, 511–522. [CrossRef]
18. Burns, R.S. The use of artificial neural networks for the intelligent optimal control of surface ships. *Ocean Eng.* **1995**, *20*, 65–72. [CrossRef]
19. Dai, S.; Wang, C.; Luo, F. Identification and learning control of ocean surface ship using neural networks. *IEEE Trans. Ind. Inform.* **2012**, *8*, 801–810. [CrossRef]
20. Nicolau, V. On PID controller design by combining pole placement technique with symmetrical optimum criterion. *Math. Probl. Eng.* **2013**, *2013*, 1024–1231. [CrossRef]
21. Rigatos, G.; Tzafestas, S. Adaptive fuzzy control for the ship steering problem. *Mechatronics* **2006**, *16*, 479–489. [CrossRef]
22. Godbole, A.A.; Libin, T.R.; Talole, S.E. Extended state observer based robust pitch autopilot design for tactical missiles. *Mech. Eng. Part G J. Aerosp. Eng.* **2011**, *226*, 1482–1501. [CrossRef]
23. Wang, H.B.; Veremey, E.I.; Xue, Y. A method of the guaranteeing of optimization for the dynamics of ship moving in wave. *Vestn. Sankt-Peterburgskogo Univ. Ser. 10 Prikl. Mat. Inform. Protsessy Upravieniya* **2017**, *13*, 354–364. [CrossRef]
24. Sun, Z.P.; Wu, Q.; Li, X.G.; Wang, H.B. A design of based on the feedback linearization optimal heading control algorithm. *Inf. Technol. Res.* **2019**, *12*, 133–148. [CrossRef]
25. Veremey, E.I.; Pogozhev, S.V.; Sotnikova, M.V. Marine autopilots' multipurpose control laws synthesis for actuators time delay. *J. Mar. Sci. Eng.* **2020**, *8*, 477. [CrossRef]
26. Du, J.L.; Hu, X.; Sun, Y.Q. Adaptive robust nonlinear control design for course tracking of ships subject to external disturbances and input saturation. *IEEE Trans. Syst. Man Cybern. Syst.* **2020**, *50*, 193–202. [CrossRef]
27. Wu, R.; Du, J.L. Adaptive robust course-tracking control of time-varying uncertain ships with disturbances. *Int. J. Control Autom. Syst.* **2019**, *17*, 1847–1855. [CrossRef]
28. Xiang, X.B.; Yu, C.Y.; Lionel, L.; Zhang, J.L. Survey on fuzzy-logic-based guidance and control of marine surface vehicles and underwater vehicles. *Int. J. Fuzzy Syst.* **2018**, *20*, 572–586. [CrossRef]
29. Xu, H.T.; Fossen, T.I.; Soares, C.G. Uniformly semiglobally exponential stability of vector field guidance law and autopilot for path-following. *Eur. J. Control* **2020**, *53*, 88–97. [CrossRef]
30. Liu, Y.; Bu, R.X.; Xu, H.J. Straight-path tracking control of underactuated ship based on backstepping method. In Proceedings of the 2015 Ninth International Conference on Frontier of Computer Science and Technology FCST, Dalian, China, 26–28 August 2015; pp. 223–227.
31. Zhao, Y.; Dong, L.L. Robust path-following control of a container ship based on Serret-Frenet frame transformation. *J. Mar. Sci. Technol.* **2020**, *25*, 69–80. [CrossRef]
32. Xu, H.T.; Oliveira, P.; Guedes, S.C. L1 adaptive backstepping control for path-following of underactuated marine surface ships. *Eur. J. Control* **2020**. [CrossRef]
33. International Electrotechnical Commission. *Maritime Navigation and Radiocommunication Equipment and Systems—Track Control Systems—Operational and Performance Requirements, Methods of Testing and Required Test Results*, 2nd ed.; 62065:2014; IEC: Geneva, Switzerland, 2014; pp. 1–212.
34. Ramos, R.L. Linear quadratic optimal control of a spar-type floating offshore wind turbine in the presence of turbulent wind and different sea states. *J. Mar. Sci. Eng.* **2018**, *6*, 151. [CrossRef]
35. Das, S.; Talole, S.E. Robust steering autopilot design for marine surface vessels. *IEEE J. Ocean. Eng.* **2016**, *41*, 913–922. [CrossRef]
36. Hiadsi, S.; Bouafia, H.; Sahli, B.; Abidri, B.; Bouaza, A.; Akriche, A. Ship course identification model based on recursive least squares algorithm with dynamic forgetting factor. *J. Comput. Appl.* **2018**, *38*, 900–904.
37. Aoki, M. Control of large-scale dynamic systems by aggregation. *IEEE Trans. Autom. Control* **1986**, *13*, 246–253. [CrossRef]
38. Abedi, F.; Leong, W.J.; Abedi, M. Lyapunov characterization for the stability of stochastic control systems. *Math. Probl. Eng.* **2015**, *2015*, 1–7. [CrossRef]
39. Wang, X.F.; Zou, Z.J.; Li, T.S.; Luo, W.L. Adaptive path following controller of underactuated ships using serret-frenet frame. *J. Shanghai Jiaotong Univ.* **2010**, *15*, 334–339. [CrossRef]

Journal of
Marine Science and Engineering

MDPI

Article

L2-Gain Based Adaptive Robust Heel/Roll Reduction Control Using Fin Stabilizer during Ship Turns

Zhang Songtao [1,*] and Zhao Peng [2,*]

[1] College of Intelligent System Science and Engineering, Harbin Engineering University, Harbin 150001, China
[2] Department of Transportation and Vehicle Engineering, Tangshan University, Tangshan 063000, China
* Correspondence: hrbzst@126.com (Z.S.); tyjtxzp@163.com (Z.P.); Tel.: +86-13946032279 (Z.S.); +86-18733340204 (Z.P.)

Abstract: The rolling and heeling experienced by a ship during turning will be more severe under the interference of winds and waves, which will seriously affect the navigation safety of the ship. The fin stabilizer is currently the best active anti-rolling device, which is usually used to reduce the roll of the ship during straight-line sailing. The purpose of this work is to study the use of fin stabilizers to reduce the rolling and heeling during ship turning, considering the non-linearity and uncertainty during the rotation. The 4 degrees of freedom (4-DOF) nonlinear motion model of a multi-purpose naval vessel is established. The forces and moments produced by fin stabilizers, rudders, propellers, and waves are also considered. The nonlinear control model of rotation and roll is derived and established. Given the non-linearity and uncertainty in the ship turning process, an L2-gain based robust adaptive control is proposed to control the fin stabilizers to reduce the turning heel and roll motion. The proof of the stability and the detailed design process of the controller are also given. Simulations are carried out to verify the effectiveness of the proposed control strategy. For comparison purposes, the simulation results under a well-tuned PID controller are also given. The simulation results show that the developed control strategy can effectively reduce the heel and roll during ship turns, and it has good robustness against uncertainty and internal and external interference.

Keywords: fin stabilizer; ship turning; heel/roll reduction; L2-gain; uncertainty; non-linearity

Citation: Songtao, Z.; Peng, Z. L2-Gain Based Adaptive Robust Heel/Roll Reduction Control Using Fin Stabilizer during Ship Turns. *J. Mar. Sci. Eng.* **2021**, *9*, 89. https://doi.org/10.3390/jmse9010089

Received: 19 December 2020
Accepted: 12 January 2021
Published: 15 January 2021

1. Introduction

When a ship is sailing at sea, it will experience movement in six degrees of freedom under the influence of winds, waves, and currents [1]. The roll motion, mainly caused by waves, has the largest impact on ship navigation safety [2]. Researchers and engineers have designed and manufactured a variety of passive or active anti-rolling devices to reduce ship rolling, such as bilge keel [1], anti-rolling tanks [3], moving weights [4], Magnus rotating roll stabilizers [5], and fin stabilizers [1,6,7]. The rudder can also be used to reduce the roll of the ship while maintaining or changing the course [8–10]. Among them, the fin stabilizer is currently the most effective active roll reduction device in the world [11].

In the traditional sense, the design and optimization of fin stabilizers are all aimed at minimizing the rolling motion of the ship caused by disturbances, such as sea waves during regular straight-line navigation, to ensure that the ship can sail smoothly and safely [6,12]. However, in some cases, ships also need to be steered for turning. The ship will also undergo a certain amount of rolling and heeling motions during turning in calm water [10]. For military ships, in order to avoid dangers, such as incoming torpedoes, emergency maneuvers with large rudder angles are required at high speeds [12]. At this time, the rolling and heeling motions experienced by the ship will be quite large. If there are certain disturbances, such as waves during the turning process, the rolling motion of the ship will be more severe, which will seriously affect the navigation safety of the ship. At the same time, the large-angle heeling will also affect the normal use of shipboard radar and weapons, and affect the normal work of ship personnel. This paper mainly studies the influence of fin stabilizers on the turning performance of ships and discusses the use of fin stabilizers to reduce the ship's heeling and rolling during high-speed and large rudder angle turning.

Therefore, it is of more practical significance to use existing anti-rolling devices such as fin stabilizers to reduce roll and heel during ship turning. The French Charles de Gaulle aircraft carrier installed the Cogite system under the flight deck to reduce the heel during emergency turning to ensure a stable take-off and landing environment for the carrier aircraft [13]. Martin proposed a three-step control strategy to control the fin stabilizer to achieve the purpose of controlling heel and turning circle [12]. Wang designed a neuron adaptive PID controller to control the fin stabilizer to realize the roll reduction control during ship turns [14]. Zhang studied the influence of passive anti-rolling tanks on ship turning characteristics [15]. Liang et al. studied the influence of rudder type parameters on ship turning [16]. Liu proposed PID control based on particle swarm optimization to reduce the rolling motion during ship turning [17]. Liang et al. Developed an improved PID controller based on multi-island genetic algorithm to reduce the roll during ship turns [18]. The fin stabilizers of most ships are controlled by a PID controller developed using classical control theory [2,19]. The control parameters are usually designed and obtained using a simplified model for a limited number of environment conditions [19,20]. Uncertain parameters and environmental disturbances will decrease the control effect. Perez and Blanke pointed out that one of the key issues in roll reduction control is the adaptation to the changes in the environmental conditions [6]. Compared with the straight-line heading state, the nonlinearities and uncertainties of the ship in the turning process are more significant. For such a nonlinear system with uncertainty, it is difficult for conventional linear methods to achieve ideal control results.

So far, the control research on uncertain nonlinear systems can be divided into three categories: one is to use adaptive control methods to solve the uncertainty of non-matching conditions of nonlinear systems; the other is to use robust control methods to ensure the internal stability and disturbance attenuation of the closed-loop system; the third is a combination of the first two methods, that is, the adaptive robust control method. All of them can enhance the robustness of nonlinear systems to uncertainties, so they have attracted the attention of experts and scholars in the field of ship motion control with extensive uncertainty and non-linearity. Zhang et al. discussed the application of L2-gain robust adaptive control in force control fin stabilizer system [21]. Wang and Zhang designed the rudder-fin joint nonlinear robust controller based on backstepping and closed-loop gain shaping method and achieved good results through simulation [22]. Li et al. designed a robust adaptive control strategy for underwater remote-controlled submersibles with speed constraints [23]. Kahveci and Ioannou proposed an adaptive steering control method including linear quadratic controller and anti-saturation compensator based on Riccati for uncertain ship dynamics affected by input constraints [24]. Zhang et al. proposed a control scheme based on a simplified adaptive neural network to solve the problem of adaptive path following control of under-driven ships with model uncertainty and non-zero mean time-varying disturbances [25]. Considering the modeling error and the uncertainty of environmental interference, Hinostroza et al. proposed a robust fin controller based on L2-gain to reduce the rolling motion of surface ships [19]. Demirel and Alarcin designed H2 and H∞ state feedback controllers based on linear matrix inequalities for fin stabilization for the effects of roll nonlinearity and uncertainty [26]. Zwierzewice uses a robust adaptive feedback linearization method to complete the design of the ship's automatic steering instrument controller [27]. Sun et al. proposed a nonlinear robust adaptive scheme based on PI sliding mode control and interference upper limit estimation to improve the robustness of the under-driven surface ship motion control system with model uncertainty and environmental disturbance [28]. Aiming at the uncertainty and environmental disturbance, in this paper, an adaptive robust control method based on

L2-gain is adopted to design the fin controller to reduce the heeling and rolling during ship turns.

The structure of the paper is as follows. Section 2 establishes the mathematical models of the ship motion and the forces and moments acting on the hull. Section 3 discusses the effect of fin stabilizers on ship turning motion. Section 4 obtains the control model and designs the L2-gain adaptive robust controller. Section 5 gives the simulation results and discussion. Finally, the conclusion is given.

2. Ship Motion Model

2.1. 4-DOF Motion Model

In order to accurately describe the motion of the ship with six degrees of freedom, the inertial coordinate system and the body-fixed coordinate system are established, as shown in Figure 1.

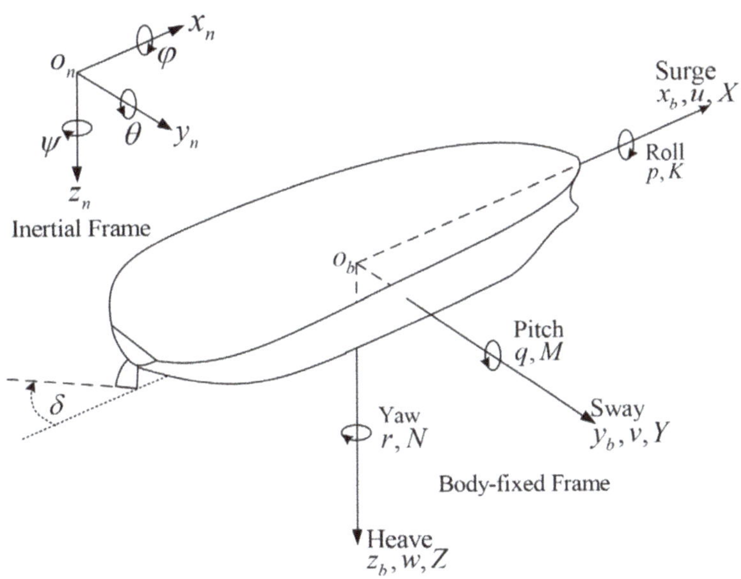

Figure 1. Coordinate system for ship modeling.

Regarding the ship as a rigid body, the 6-DOF nonlinear equations of ship motion can be obtained by Newton's law [29]. Since this paper mainly studies the use of fin stabilizer to reduce the heeling and rolling motion caused by ship turning and environmental disturbance, the pitch and heave attitudes, and their influence on the other four degrees of freedom can be neglected [30]. Therefore, the 4-DOF ship motion model in the surge, sway, roll and yaw can be obtained by Equation (1):

$$\begin{aligned} m(\dot{u} - vr - x_G r^2 + z_G pr) &= X \\ m(\dot{v} + ur - z_G \dot{p} + x_G \dot{r}) &= Y \\ I_x \dot{p} - m z_G (\dot{v} + ur) &= K \\ I_z \dot{r} + m x_G (\dot{v} + ur) &= N \end{aligned} \tag{1}$$

where m is the ship mass, x_G and z_G are the coordinates of the center of gravity (CG) in x and z axes, respectively, I_x and I_z are the moments of inertia about the x and z axes, u and v are the surge and sway velocity, respectively, p and r are the roll and yaw angular velocity, X and Y are the forces acting on the hull in x and y axes direction, respectively, K and N are the moments acting on the hull about x and z axes, respectively.

2.2. Hydrodynamic Forces and Moments

The hydrodynamic forces and moments can be expressed in the form of nonlinear functions of the ship's position, velocity and acceleration, and are influenced by the shipping state and fluid characteristics [29]. In this paper, the nonlinear hydrodynamic expression is adopted as [1]:

$$\begin{aligned}
X_{hyd} = \; & X_{\dot{u}}\dot{u} + X_{u|u|}u|u| + X_{vr}vr \\
Y_{hyd} = \; & Y_{\dot{v}}\dot{v} + Y_{\dot{p}}\dot{p} + Y_{\dot{r}}\dot{r} + Y_{|u|v}|u|v + Y_{ur}ur + Y_{v|v|}v|v| + Y_{v|r|}v|r| + Y_{r|v|}r|v| + \\
& Y_{\varphi|uv|}\varphi|uv| + Y_{\varphi|ur|}\varphi|ur| + Y_{\varphi uu}\varphi uu \\
K_{hyd} = \; & K_{\dot{v}}\dot{v} + K_{\dot{p}}\dot{p} + K_{|u|v}|u|v + K_{ur}ur + K_{v|v|}v|v| + K_{v|r|}v|r| + K_{r|v|}r|v| + \\
& K_{\varphi|uv|}\varphi|uv| + K_{\varphi|ur|}\varphi|ur| + K_{\varphi uu}\varphi uu + K_{|u|p}|u|p + K_{|p|p}|p|p + K_{p}p + \\
& K_{\varphi\varphi\varphi}\varphi\varphi\varphi - \rho g \nabla GZ(\varphi) \\
N_{hyd} = \; & N_{\dot{v}}\dot{v} + N_{\dot{r}}\dot{r} + N_{|u|v}|u|v + N_{|u|r}|u|r + N_{r|r|}r|r| + N_{r|v|}r|v| + N_{\varphi|uv|}\varphi|uv| + \\
& N_{\varphi u|r|}\varphi u|r| + N_{|p|p}|p|p + N_{p}p + N_{|u|p}|u|p + N_{\varphi u|u|}\varphi u|u|
\end{aligned} \tag{2}$$

where ρ is the fluid density, g is the gravity acceleration, ∇ is the ship displacement, φ is the roll angle, $GZ(\varphi)$ is the righting arm curve, and the multipliers are the ship's hydrodynamic coefficients.

2.3. Wave Disturbance Forces and Moments

Disturbances experienced by ships sailing at sea are mainly caused by sea winds, waves, and currents. Among them, sea waves are the main cause of ship rolling. Considering the research content of this paper, the disturbance forces and moments produced by sea waves are considered, and the wave forces and moments model is adopted as [31]:

$$\begin{aligned}
X_{wave} &= -\rho g \cos\chi \sum_{n=1}^{NN} E_n \frac{\omega_n^2}{g} [A_n \cos(\omega_e t + \varepsilon_n) - B_n \sin(\omega_e t + \varepsilon_n)] \\
Y_{wave} &= \rho g \sin\chi \sum_{n=1}^{NN} E_n \frac{\omega_n^2}{g} [A_n \cos(\omega_e t + \varepsilon_n) - B_n \sin(\omega_e t + \varepsilon_n)] \\
K_{wave} &= -\rho g \sin\chi \sum_{n=1}^{NN} E_n \frac{\omega_n^2}{g} [C_n \cos(\omega_e t + \varepsilon_n) - D_n \sin(\omega_e t + \varepsilon_n)] \\
N_{wave} &= \rho g \sin\chi \sum_{n=1}^{NN} E_n \frac{\omega_n^2}{g} [G_n \cos(\omega_e t + \varepsilon_n) - H_n \sin(\omega_e t + \varepsilon_n)]
\end{aligned} \tag{3}$$

where

$$A_n = \int F_n(x)A(x)\sin\left(\frac{\omega_n^2}{g}x\cos\chi\right)dx, \; B_n = \int F_n(x)A(x)\cos\left(\frac{\omega_n^2}{g}x\cos\chi\right)dx$$

$$C_n = \int F_n(x)z_{B(x)}A(x)\sin\left(\frac{\omega_n^2}{g}x\cos\chi\right)dx, \; D_n = \int F_n(x)z_{B(x)}A(x)\cos\left(\frac{\omega_n^2}{g}x\cos\chi\right)dx$$

$$G_n = \int F_n(x)xA(x)\sin\left(\frac{\omega_n^2}{g}x\cos\chi\right)dx, \; H_n = \int F_n(x)xA(x)\cos\left(\frac{\omega_n^2}{g}x\cos\chi\right)dx$$

$$F_n = \sin\left(\frac{\omega_n^2}{g}\frac{B(x)}{2}\sin\chi\right)e^{\frac{\omega_n^2}{g}d(x)} / \frac{\omega_n^2}{g}\frac{B(x)}{2}\sin\chi, \; E_n = \sqrt{2S_\zeta(\omega_n)\Delta\omega}$$

where A_n, B_n, C_n, D_n, G_n and H_n are the corresponding coefficients of the first-order wave disturbance forces and moments. $A(x)$, $B(x)$, $d(x)$ and $z_{B(x)}$ are the cross-section area of water immersion, beam, draft and the z coordinate of the center of the cross-section area of water immersion at the ship's longitudinal coordinate of x, respectively. ω_n and ω_e are the wave frequency and encounter frequency, respectively. χ is the encounter angle. ε_n and $S_{\zeta(\omega n)}$ are the random phase and wave-height spectrum of the nth regular wave, respectively. NN is the number of regular waves.

2.4. Fin Forces and Moments

Fin stabilizers are installed in pairs on both sides of the bilge of the ship to have a large roll arm and anti-rolling moment. According to [1], the fin forces and moments can be approximated as:

$$\begin{aligned} X_{fin} &\approx -2T_f \\ Y_{fin} &\approx 2N_f \cdot \sin\left(\beta_{fin}\right) \\ K_{fin} &\approx -2r_f \cdot N_f \\ N_{fin} &\approx -2FCG \cdot N_f \cdot \sin\left(\beta_{fin}\right) \end{aligned} \tag{4}$$

where r_f and FCG are the fin's roll and yaw arms, respectively, β_{fin} is the fin's tilt angle, T_f and N_f are the fin-induced tangential and normal forces, respectively, and can be calculated by Equation (5):

$$\begin{aligned} N_f &= D_f \sin\alpha_f + L_f \cos\alpha_f \\ T_f &= D_f \cos\alpha_f - L_f \sin\alpha_f \end{aligned} \tag{5}$$

where L_f and D_f are the lift and drag forces generated on the fin, respectively, α_f is the effective fin angle of attack.

2.5. Rudder Forces and Moments

The vast majority of ships rely on one or more vertical installed rudders to accomplish all the maneuvers [12]. Rudders are usually installed after the propeller below the stern line of the ship. According to [14], the rudder-induced forces and moments can be obtained, as shown in Equation (6)

$$\begin{aligned} X_{rud} &= -D_r \cdot \delta \\ Y_{rud} &= L_r \cdot \delta \\ K_{rud} &= -r_r \cdot L_r \cdot \delta \\ N_{rud} &= -LCG \cdot L_r \cdot \delta \end{aligned} \tag{6}$$

where r_r and LCG are the rudder's roll and yaw arms, respectively, δ is the fin tilt angle, L_r and D_r are the lift and drag forces generated on the fin, respectively.

2.6. Propeller Forces and Moments

Most ships rely on propellers under the waterline of the stern to provide forward power. Under normal conditions, it can be considered that the propellers only produce a longitudinal force that drives the ship forward or backwards. When the ship sails straight at a constant speed, it can be considered that the effective thrust T_e generated by the propeller is equal to the ship resistance X_R. Perez suggested that the ship resistance can be considered to be approximately equal to $X_{|u|u}U^2$, where U is the ship sailing speed [1]. Martin pointed out through simulation that for ship turning at a fixed speed, the thrust generated by the propellers during the turning process can be approximately equal to the force before the turning and remains unchanged to simplify the study [12]. Therefore, according to [12], the forces and moments can be obtained, as shown in Equation (7):

$$\begin{aligned} X_{prop} &= -X_{|u|u}U_0^2 \\ Y_{prop} &= 0 \\ K_{prop} &= 0 \\ N_{prop} &= 0 \end{aligned} \tag{7}$$

where $X_{|u|u}$ is the hydrodynamic coefficient related to the surge velocity, and U_0 is the initial speed before turning.

3. Ship Heel/Roll Control Using Fin

A multipurpose naval vessel with fin stabilizers and rudders is selected as the research object, the parameters of the vessel, fin and rudder are shown in Tables 1 and 2, respectively. The fin angle that rolls the ship to port is defined as the positive fin angle. The rudder angle that turns the ship left is defined as the positive rudder angle. The seawater density is 1025 kg/m^3. The hydrodynamic coefficients presented in Section 2.2 are shown in Table 3, where m is the mass of the ship and Δ is the buoyancy of the ship.

Table 1. Parameters of the vessel.

Description	Value	Unit
Length between perpendiculars	51.5	m
Beam overall	8.6	m
Draft	2.3	m
Displacement	357	m^3
Transverse metacentric height	1.1	m
Roll inertia	2.38×10^6	kg·m^2
Yaw inertia	4.79×10^7	kg·m^2
Coordinates of center of gravity	$(-3.38, 0, -1.06)$	m

Table 2. Parameters of the fin and rudder.

Description	Fin	Rudder	Unit
Area	1.6×2	1.5×2	m^2
Aspect ratio	1	1.5	–
Max. angle of attack	25	35	°
Max. turning rate	15	10	°/s
Roll arm	4.22	3.3	m
Yaw arm	−2	20.4	m
Lift coefficient	0.042	0.044	/°
Tilt angle	34	90	°

Table 3. Hydrodynamic coefficients (from [1]).

Subscript	X	Y	K	N
$\dot{u}$	-1.74×10^4	0	0	0
$\dot{v}$	0	-1.90×10^6	2.96×10^5	5.38×10^5
$\dot{p}$	0	-2.96×10^5	-6.74×10^5	0
$\dot{r}$	0	-1.40×10^6	0	-4.40×10^7
$u\|u\|$	-1.96×10^3	0	0	0
$\|u\|v$	0	-1.18×10^4	9.26×10^3	-9.20×10^4
$\|u\|r$	0	1.31×10^5	-1.02×10^5	-4.71×10^6
$v\|v\|$	0	-3.70×10^3	2.93×10^4	0
$\|r\|r$	0	0	0	-2.02×10^8
$\|v\|r$	$0.33 \times m$	-7.94×10^5	6.21×10^5	0
$r\|v\|$	0	-1.82×10^5	1.42×10^5	-1.56×10^7
$\varphi\|uv\|$	0	-1.08×10^4	-8.40×10^3	-2.14×10^5
$\varphi\|ur\|$	0	-2.51×10^5	-1.96×10^5	-4.98×10^6
φuu	0	-7.40×10^1	-1.18×10^3	-8.00×10^3
$\|u\|p$	0	0	-1.55×10^4	0
$\|p\|p$	0	0	-4.16×10^5	0
p	0	0	-5.00×10^5	0
$\varphi\varphi\varphi$	0	0	-0.325Δ	0

The correctness and applicability of the established model have been verified in our previous work [10]. Figure 2 shows the simulation results of ship turning motion under different sea states. The simulation parameters are as follows: the initial ship speed is 15 m/s, the fin angle is 0°, the rudder angle is 30°, the rudder rate is 10°/s, the initial encounter angle is 90°.

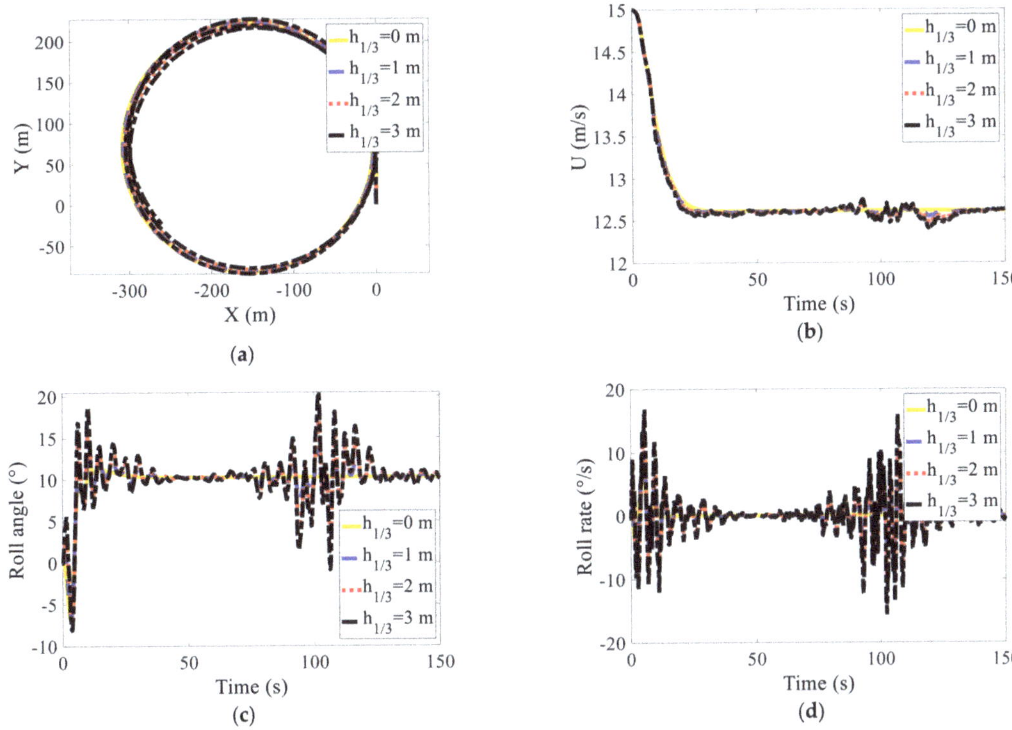

Figure 2. Simulation results of ship turning motion under different sea states. (**a**) Trajectory; (**b**) ship speed; (**c**) roll angle; (**d**) roll rate.

It can be seen from Figure 2 that the turning trajectory of the ship under the interference of sea waves has a slight deviation compared to that in calm water, and the turning diameter is basically unchanged. The outward heel and ship speed in the steady turning phase also changed irregularly on the basis of the steady outward heel angle and the steady sailing speed of the turning in clam water, respectively. As the level of sea conditions increases, the changes in the state of motion, especially the turning heel, are also intensified. Under the interference of the waves with a significant wave height of 3 m, the turning outward heel angle reaches a maximum of 20°, which seriously affects the navigation safety and comfort of the ship. Therefore, the ship should try to avoid turning manoeuvres in higher sea conditions. The sea state level of the subsequent heel/roll reduction control simulations in this paper is limited to sea state 3 with a significant wave height of 1.25 m.

3.1. Ship Turning Motion with Fin Stabilizer

To investigate the effect of fin stabilizer on ship turning motion, the simulation of the ship turning in calm water with different fin angles of attack is conducted. The initial ship speed is 15 m/s, the rudder angle of attack is 30° and the rudder turning rate is 10°/s. The simulation results with fin angle of −20°, −10°, 0°, 10°, and 20° are shown in Figure 3. The main results data are given in Table 4. It should be noted that the total simulation time is 100 s, and the simulation results of the first 50 s of the roll angular velocity are given to clearly show the changes in the transition process.

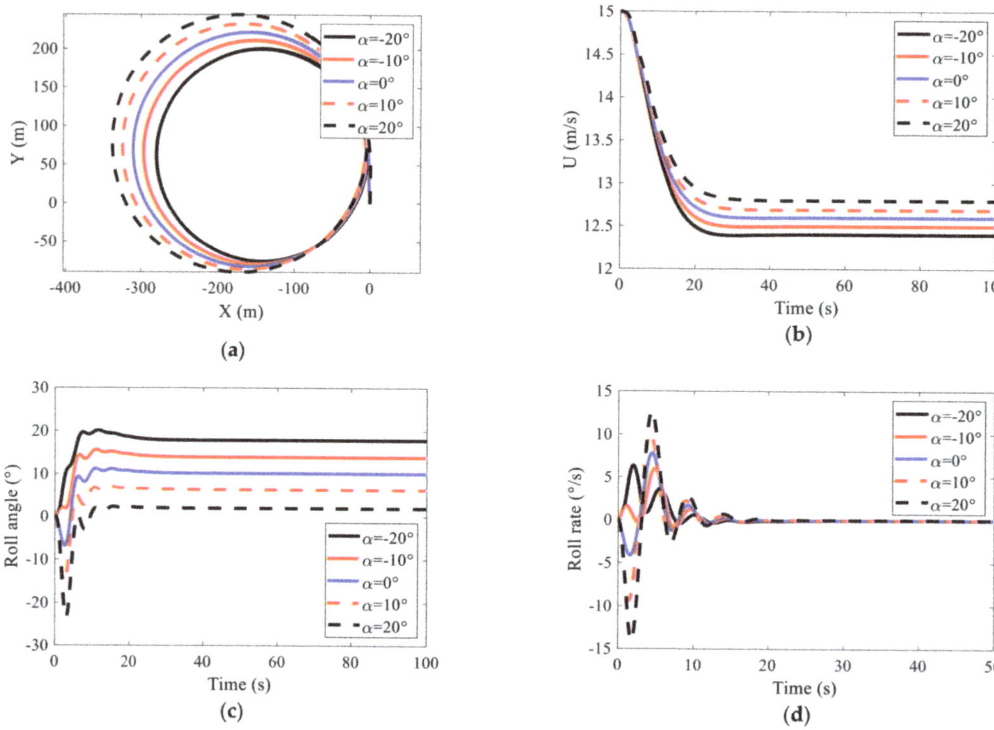

Figure 3. Simulation results of ship turning motion under different fin angles. (**a**) Trajectory; (**b**) ship speed; (**c**) roll angle; (**d**) roll rate.

Table 4. Simulation results of turning circle parameters under different fin angles.

Fin Angle (°)	Advance (m)	Steady Turning Diameter (m)	Max. Inward/Outward Heel Angle (°)	Steady Outward Heel Angle (°)	Steady Sailing Speed (m/s)
−20	198.63	276.29	0/20.14	17.86	12.41
−10	208.98	290.68	0/15.67	13.96	12.50
0	218.92	304.53	−6.79/11.29	10.22	12.60
10	229.45	318.57	−15.29/7.01	6.30	12.69
20	241.05	334.41	−23.36/2.30	1.91	12.80

It can be seen from Figure 3 and Table 4 that the positive and negative fin angles have opposite effects on the ship's turning performance, and the strength of the influence increases as the fin angle increases. Under the action of a positive fin angle, the advance, turning radius and initial inward heel are increased compared to when the rudders act alone, while the maximum and steady outward heel angles are reduced. When the negative fin angle is acting, the opposite is true. In addition to the relatively small impact of the fin stabilizer on the speed reduction of turning, it also has a greater impact on other parameters such as the advance, turning diameter and turning heel. Therefore, the fin stabilizer can be used to assist the rudder to change the ship's turning characteristics. In addition, it can be seen from the simulation results that the influence of the fin stabilizer on the steady turning diameter and steady outward heel is exactly the opposite. If one is reduced, the other is bound to increase. It is necessary to adjust the corresponding weight coefficients according to different control objectives to achieve relatively optimal control effects.

3.2. Fin Angle Speed Adjustment

It can be seen from the simulation results in Section 3.1 that fin stabilizers can be used to reduce the heel and roll motion during ship turns. It should be noted that to explore the law, the fin angle speed adjustment is not considered when setting the fin angle in the simulation of the front section. In the actual navigation of the ship, to protect the fin shaft from damage at high speeds, the speed adjustment problem must be considered when using fin stabilizers to reduce roll, which is, adjusting the maximum angle of attack of the fin according to the speed. In order to make the simulation results closer to the real situation, in this paper, the following fin angle speed adjustment law is adopted [32]:

$$\alpha^*_{\max} = \begin{cases} 0 & U < U_{L1}, U < U_{L2} \\ \alpha_{\max} & U_{L1} \le U \le U_{design}, U_{L2} \le U \le U_{design} \\ \left(U_{design}/U\right)^2 \alpha_{\max} & U > U_{design} \end{cases} \tag{8}$$

where $\alpha^*_{\max}$ is the real-time maximum fin angle, $\alpha_{\max}$ is the maximum fin angle corresponding to the design speed, U_{design} is the design speed. In this paper, U_{design} and $\alpha_{\max}$ are set to 18 kn and 25°, respectively.

4. Heel/Roll Reduction Control Using Fin Stabilizers

Based on the above discussion, the diagram of the heel/roll reduction control system using fin stabilizers during ship turns, as shown in Figure 4, can be obtained.

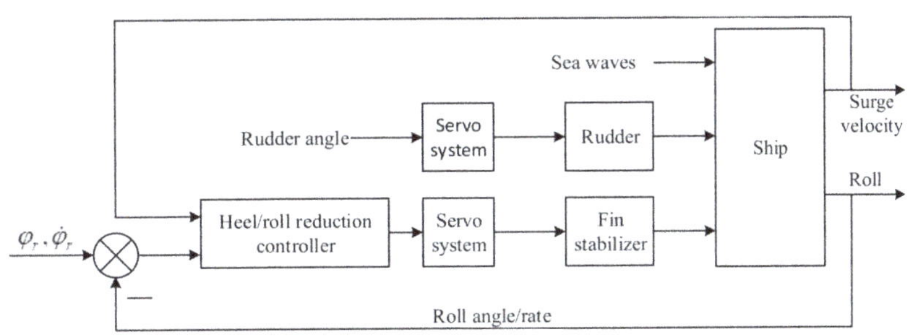

Figure 4. Diagram of heel/roll reduction control using fin stabilizers during ship turns.

4.1. Control Model Analysis

As mentioned earlier, when using fin stabilizers to reduce the heel during the turning process, it will inevitably increase the turning diameter of the ship. Therefore, in order to simplify the analysis, only one degree of freedom of roll is considered. From the mathematical model of ship motion established in Section 2, the ship's roll model is given by Equation (9):

$$\begin{aligned} I_x\dot{p} - mz_G(\dot{v} + ur) = {} & K_{\dot{v}}\dot{v} + K_{\dot{p}}\dot{p} + K_{|u|v}|u|v + K_{ur}ur + K_{|v|v}|v|v + K_{|v|r}|v|r + \\ & K_{\phi|uv|}\phi|uv| + K_{\phi|ur|}\phi|ur| + K_{\phi uu}\phi uu + K_{|u|p}|u|p + K_{|p|p}|p|p + \\ & K_p p + K_{\phi\phi\phi}\phi\phi\phi - \rho g\nabla GMt\sin(\phi) + K_{rud} + K_{fin} + K_w \end{aligned} \tag{9}$$

Liang et al. pointed out that the sway acceleration term has little effect on the roll, and can be neglected [18]. In addition, when $\varphi < 28°$, $\sin(\varphi)$ can be approximately regarded as equal to φ. Therefore, ignoring the sway acceleration term and rewriting K_{fin} as $K_{f\alpha f}$, Equation (9) can be rewritten as:

$$\begin{aligned} \left(I_x - K_{\dot{p}}\right)\dot{p} = {} & \left(K_{|u|p}|u| + K_p\right)p + K_{|p|p}|p|p + \\ & \left(K_{\phi|uv|}|uv| + K_{\phi|ur|}|ur| + K_{\phi uu}uu - \rho g\nabla GMt\right)\phi + K_{\phi\phi\phi}\phi\phi\phi + \\ & K_{|u|v}|u|v + K_{ur}ur + K_{|v|v}|v|v + K_{|v|r}|v|r + K_{|r|v}|r|v + mz_G ur + K_{rud} + K_f\alpha_f + K_w \end{aligned} \tag{10}$$

where $K_f = \rho A_f u^2 C_L r_f$, C_L is the lift coefficient of the fin and A_f is the fin area.

Define $v = \alpha_f$, and Equation (10) can be further simplified as:

$$\dot{p} = \theta_1 p + \theta_2 |p| p + \theta_3 \phi + \theta_4 \phi^3 + bv + f_w \tag{11}$$

where

$$\theta_1 = \left(K_{|u|p}|u| + K_p\right) / \left(I_x - K_{\dot{p}}\right),\ \theta_2 = K_{|p|p} / \left(I_x - K_{\dot{p}}\right)$$

$$\theta_3 = \left(K_{\phi|uv|}|uv| + K_{\phi|ur|}|ur| + K_{\phi uu}uu - \rho g \nabla GMt\right) / \left(I_x - K_{\dot{p}}\right)$$

$$b = K_f / \left(I_x - K_{\dot{p}}\right)$$

$$f_w = \left(K_{|u|v}|u|v + K_{ur}ur + K_{|v|v}|v|v + K_{|v|r}|v|r + mz_G ur + K_{rud} + K_W\right) / \left(I_x - K_{\dot{p}}\right)$$

It can be seen that the equation describing the roll motion in the turning process is strongly nonlinear, and the parameters of the equation still have greater uncertainty due to the speed reduction of turning and the influence of external interference. In this paper, the L2-gain based adaptive robust control method is adopted to design the fin controller to reduce the turning heel and roll motion.

4.2. Controller Design

4.2.1. L2-Gain Adaptive Robust Control

For the disturbed uncertain nonlinear system described in Equation (12), suppose it satisfies the following assumptions in a certain neighborhood of the origin of the state space [33].

$$\begin{cases} \dot{x}_i = x_{i+1} + \gamma_i^T(x_1, x_2, \cdots, x_i), 1 \le i \le n-1 \\ \dot{x}_n = \gamma_0(x) + \gamma_n^T(x)\theta + \left(\beta_0(x) - \beta^T(x)\theta\right)v + \eta(x,t) \\ y = x_1 \end{cases} \tag{12}$$

where $x \in R^n$, $v \in R$, $y \in R$, and $\eta \in R$ are the state vector, input, output and bounded disturbance of the system, respectively, $\gamma_i(x)$ and $\beta(x)$ are smooth vector fields with $0 \le i \le n$, $\theta = (\theta 1, \theta 2, \ldots, \theta p)^T \in L_2(0,\infty)$ is the uncertain disturbance.

Hypothesis 1: (assumption of bounded interference) there is a constant d greater than zero, satisfying $|\eta(x,t)| \le d$;

Hypothesis 2: (achievable condition hypothesis) let $\hat{\theta}$ be the estimate of θ, for any $\hat{\theta} \in B_\theta$ and $x \in B_x$, it needs to satisfy $|\beta_0(x) + \beta^T(x)\hat{\theta}| > 0$, where B_θ and B_x are the open balls that contain $\hat{\theta}$ and x respectively in the parameter space. Compared with the control period, $\theta \in L_2(0,\infty)$ is usually regarded as a slowly changing quantity, so it can be approximately equal to zero.

Then, introduce the following theorem as [34].

Theorem 1. *For any positive real constant $\gamma > 0$, if the perturbed nonlinear system described in Equation (12), for any $T \ge 0$ and $\theta \in L_2(0,T)$, satisfy.*

$$\int_0^T \|y(t)\|^2 dt \le \gamma^2 \int_0^T \|\theta\|^2 dt + N \tag{13}$$

where $N \ge 0$ is a finite constant, then the disturbing uncertain nonlinear system Equation (12) has L2-gain $\le \gamma$.

The design steps of L2-gain adaptive robust controller based on the backstepping method are briefly given below. For succinct description, let $\gamma_i(x1, x2, \ldots, xi)$ be γ_i, then when $i = 1$, we can get from Equation (12):

$$\dot{x}_1 = x_2 + \gamma_1^T \theta \tag{14}$$

Assume $x_2{}^*$ is the control input of the subsystem (14), and defined by Equation (15)

$$x_2^* = -c_1 x_1 - \frac{1}{4} k x_1 \left(1 + \gamma_1^T \gamma_1\right) \tag{15}$$

where $c_1 > 0$ and $k > 0$ are undetermined coefficients.

Define the error term $e_2 = x_2 - x_2{}^*$, then

$$x_2 = e_2 + x_2^* = e_2 - c_1 x_1 - \frac{1}{4} k x_1 \left(1 + \gamma_1^T \gamma_1\right) \tag{16}$$

Define the Lyapunov function V_1 as

$$V_1 = \frac{1}{2} x_1^2 \tag{17}$$

Then we have Equation (18)

$$\dot{V}_1 = x_1 \dot{x}_1 = x_1 \left(x_2 + \gamma_1^T \theta\right) \tag{18}$$

Substitute Equation (16) into (18), we can get Equation (19)

$$\dot{V}_1 = x_1 e_2 - c_1 x_1^2 - \frac{1}{4} k x_1^2 \left(1 + \gamma_1^T \gamma_1\right) + x_1 \gamma_1^T \theta \tag{19}$$

The last two terms on the right side of Equation (19) can be rewritten as

$$-\tfrac{1}{4} k x_1^2 (1 + \gamma_1^T \gamma_1) + x_1 \gamma_1^T \theta = -k \left(\tfrac{1}{2} x_1 \sqrt{1 + \gamma_1^T \gamma_1} - \frac{\gamma_1^T \theta}{k \sqrt{1 + \gamma_1^T \gamma_1}} \right)^2 + \frac{(\gamma_1^T \theta)^2}{k (1 + \gamma_1^T \gamma_1)} \leq \tfrac{1}{k} \|\theta\|^2 \tag{20}$$

Let $e_1 = x_1$, then we have Equation (21)

$$\dot{V}_1 \leq -c_1 x_1^2 + \frac{1}{k} \|\theta\|^2 + e_1 e_2 \tag{21}$$

For $i = 2$, define the Lyapunov function V_2 as

$$V_2 = V_1 + \frac{1}{2} e_2^2 \tag{22}$$

Then we have Equation (23)

$$\dot{V}_2 = \dot{V}_1 + e_2 \dot{e}_2 \leq -c_1 e_1^2 + \frac{1}{k} \|\theta\|^2 + e_2 e_1 + e_2 \dot{e}_2 \tag{23}$$

From the definition of the previous error term e_2, we can get Equation (24)

$$\dot{e}_2 = \dot{x}_2 - \dot{x}_2^* = x_3 - \frac{\partial x_2^*}{\partial x_1} x_2 + \left(\gamma_2 - \frac{\partial x_2^*}{\partial x_1} \gamma_1\right)^T \theta \tag{24}$$

Substitute Equation (24) into (23), we have

$$\dot{V}_2 \leq -c_1 e_1^2 + \frac{1}{k} \|\theta\|^2 + e_2 e_1 + e_2 \left(x_3 - \frac{\partial x_2^*}{\partial x_1} x_2 + \left(\gamma_2 - \frac{\partial x_2^*}{\partial x_1} \gamma_1\right)^T \theta \right) \tag{25}$$

If α_{21} and α_{22} are

$$\alpha_{21} = e_1 - \frac{\partial x_2^*}{\partial x_1} x_2, \alpha_{22} = \gamma_2 - \frac{\partial x_2^*}{\partial x_1} \gamma_1 \tag{26}$$

Then Equation (25) can be further written as

$$\dot{V}_2 \le -c_1 e_1^2 + \frac{1}{k}\|\theta\|^2 + e_2\left(x_3 + \alpha_{21} + \alpha_{22}^T\theta\right) \tag{27}$$

Let x_3^* be the control input of the second subsystem, and define

$$x_3^* = -c_2 e_2 - \alpha_{21} - \frac{1}{4}ke_2\left(1 - \alpha_{22}^T\alpha_{22}\right) \tag{28}$$

Moreover, define the error term e_3 between x_3 and x_3^* as

$$e_3 = x_3 - x_3^* = x_3 + c_2 e_2 + \alpha_{21} + \frac{1}{4}ke_2\left(1 - \alpha_{22}^T\alpha_{22}\right) \tag{29}$$

Substituting Equation (29) into (27), and performing a calculation similar to $\dot{V}_1$, we can get

$$\dot{V}_2 \le -c_1 e_1^2 + \frac{1}{k}\|\theta\|^2 + e_2\left(e_3 + x_3^* + \alpha_{21} + \alpha_{22}^T\theta\right) \le -\sum_{i=1}^{2} c_i e_i^2 + \frac{2}{k}\|\theta\|^2 + e_2 e_3 \tag{30}$$

Similarly, for $2 < i < n-1$, define the Lyapunov function V_i as

$$V_i = V_{i-1} + \frac{1}{2}e_i^2 \tag{31}$$

If α_{i1}, α_{i2} and the error term e_{i+1} between x_{i+1} and the control input of the ith subsystem x_{i+1}^* are defined as

$$\alpha_{i1} = e_{i-1} - \sum_{j=1}^{i-1}\frac{\partial x_i^*}{\partial x_j}x_{j+1}, \alpha_{i2} = \gamma_i - \sum_{j=1}^{i-1}\frac{\partial x_i^*}{\partial x_j}\gamma_j \tag{32}$$

$$x_{i+1}^* = -c_i e_i - \alpha_{i1} - \frac{1}{4}ke_i\left(1 + \alpha_{i2}^T\alpha_{i2}\right), e_{i+1} = x_{i+1} - x_{i+1}^* \tag{33}$$

Similarly, we can get

$$\dot{V}_i \le -\sum_{j=1}^{i} c_j e_j^2 + \frac{i}{k}\|\theta\|^2 + e_i e_{i+1} \tag{34}$$

For the nth subsystem, define the error term e_n as

$$e_n = x_n - x_n^* \tag{35}$$

Then we can have

$$\dot{e}_n = \dot{x}_n - \dot{x}_n^* = \gamma_0 - \sum_{j=1}^{n-1}\frac{\partial x_n^*}{\partial x_j}x_{j+1} + \left(\gamma_n - \sum_{j=1}^{n-1}\frac{\partial x_n^*}{\partial x_j}\gamma_j\right)^T\theta + \left(\beta_0 + \beta^T\theta\right)u + \eta \tag{36}$$

Similarly, define α_{n1} and α_{n2} as

$$\alpha_{n1} = \gamma_0 - \sum_{j=1}^{n-1}\frac{\partial x_n^*}{\partial x_j}x_{j+1}, \alpha_{n2} = \gamma_n - \sum_{j=1}^{n-1}\frac{\partial x_n^*}{\partial x_j}\gamma_j \tag{37}$$

Then Equation (36) can be further written as

$$\dot{e}_n = \alpha_{n1} + \alpha_{n2}^T\theta + \left(\beta_0 + \beta^T\theta\right)u + \eta \tag{38}$$

Define the Lyapunov function V_n as

$$V_n = V_{n-1} + \frac{1}{2}e_n^2 + \frac{1}{2}(\theta - \hat{\theta})^T(\theta - \hat{\theta}) \tag{39}$$

Then we have

$$\dot{V}_n \leq -\sum_{i=1}^{n-1} c_i e_i^2 + \tfrac{n-1}{k}\|\theta\|^2 + e_n e_{n-1} + e_n(\alpha_{n1} + \alpha_{n2}^T\theta + (\beta_0 + \beta^T\theta)u + \eta) - (\theta - \hat{\theta})\dot{\hat{\theta}} \tag{40}$$

Define the adaptive control law as

$$v = \frac{1}{\beta_0 + \beta^T\theta}\left(-\alpha_{n1} - c_n e_n - \alpha_{n2}^T\hat{\theta} - \eta\right) \tag{41}$$

$$\dot{\hat{\theta}} = \frac{-(\alpha_{n1} + c_n e_n + \hat{\theta}^T\alpha_{n2} + d\cdot \mathrm{sgn}(e_n))\beta + (\beta_0 + \beta^T\hat{\theta})\alpha_{n2}}{\beta_0 + \beta^T\hat{\theta}} e_n \tag{42}$$

Substituting the above adaptive control law into Equation (40), we can get

$$\dot{V}_n \leq -\sum_{i=1}^{n-2} c_i e_i^2 + \tfrac{n-1}{k}\|\theta\|^2 + e_n e_{n-1} = -\sum_{i=1}^{n-2} c_i e_i^2 - (c_{n-1}e_{n-1}^2 + c_n e_n^2 - e_n e_{n-1}) + \tfrac{n-1}{k}\|\theta\|^2 \tag{43}$$

Equation can be further written as follows if we choose $c_i \geq 1, 1 \leq i \leq n$

$$\dot{V}_n \leq -c_1 e_1^2 + \frac{n-1}{k}\|\theta\|^2 \tag{44}$$

Moreover, because $y = x_1 = e_1$, integrating both ends of Equation (44), we can get

$$\int_0^T \|y\|^2 dt \leq \gamma^2 \int_0^T \|\theta\|^2 dt + N \tag{45}$$

where $N > 0$ is the initial value of V_n, $\gamma^2 = (n-1)/k$.

It can be seen that, under the effect of the designed adaptive control law, the disturbed uncertain nonlinear system has L2-gain $\leq \gamma$.

4.2.2. Design of Turning Heel/Roll Reduction Controller

For the disturbed and uncertain nonlinear turning heel/roll reduction system described in Equation (11), Let $x_1 = \varphi$, $x_2 = p$, then Equation (11) can be rewritten as:

$$\begin{aligned} \dot{x}_1 &= x_2 \\ \dot{x}_2 &= \theta_1 x_2 + \theta_2|x_2|x_2 + \theta_3 x_1 + \theta_4 x_1^3 + bv + f_w \end{aligned} \tag{46}$$

Comparing the disturbed uncertain nonlinear system (46) and (12), it can be seen that the order of the turning hell/roll reduction system n is 2, and we can get

$$\gamma_1(x_1) = 0, \gamma_0 = 0, \gamma_2 = \begin{bmatrix} x_2 & |x_2|x_2 & x_1 & x_1^3 \end{bmatrix}^T \tag{47}$$

$$\beta_0 = b, \beta_2 = 0, \eta = f_w \tag{48}$$

According to the design steps of the L2-gain adaptive robust law described in Section 4.2.1, we can get

$$x_2^* = -\left(c_1 + \frac{1}{4}k\right)x_1 \tag{49}$$

$$\alpha_{21} = \left(c_1 + \frac{1}{4}k\right)x_2 \tag{50}$$

$$\alpha_{22} = \begin{bmatrix} x_2 & |x_2|x_2 & x_1 & x_1^3 \end{bmatrix}^T \tag{51}$$

$$e_2 = x_2 + \left(c_1 + \frac{1}{4}k\right)x_1 \tag{52}$$

The hyperbolic tangent function *tanh(·)* is used instead of the sign function *sgn(·)* to eliminate the chattering of the control signal caused by the hard handoff. The adaptive control law can be expressed as:

$$v = \frac{1}{b}\left[-\left(c_1 + c_2 + \frac{1}{4}k\right)x_2 - c_2\left(c_1 + \frac{1}{4}k\right)x_1 - \alpha_{22}^T\hat{\theta} - d \cdot \tanh(e_2)\right] \tag{53}$$

$$\dot{\hat{\theta}} = \alpha_{22}e_2 \tag{54}$$

where $c_1 \geq 1$, $c_2 \geq 1$ and $k > 0$ are undetermined control parameters.

5. Result and Discussion

In this section, the simulation of the ship turning under wave disturbance are performed to verify the effectiveness of the designed L2-gain based adaptive robust controller. The simulation parameters are as follows: the significant wave height is 1.25 m, the initial encounter angle is 135°, the initial sailing speed is 15 m/s, the turning rudder angle is 30°. The other parameters can be found in Tables 1 and 2. It can be seen from the simulation results in Section 3.1, the surge velocity during the turning process gradually decreases from the initial sailing speed of 15 m/s to about 12.6 m/s, and changes slightly on this basis. Considering the short duration of the steering phase and the transition phase during the turning process, the parameters of the steady-state turning is used to calculate the parameters of the disturbed and uncertain nonlinear turning heel/roll reduction system. Based on the above analysis, we can get b = 0.0146 and d = 0.2786. Design $c_1 = c_2 = k = 4$, then the adaptive control law can be obtained from Equations (52)–(54):

$$u = \frac{1}{b}\left(-20x_1 - 9x_2 - \alpha_{22}^T\hat{\theta} + 0.2786sign(e_2)\right) \tag{55}$$

$$e_2 = 5x_1 + x_2 \tag{56}$$

The simulation results of turning heel/roll reduction control using fin stabilizers based on the design L2-gain based adaptive robust controller are shown in Figures 5 and 6. For comparison purposes, the simulation results a well-tuned PID controller are also given. NC, PID and L2ARC stand for "No control", "PID control" and "L2-gain based adaptive robust control" respectively.

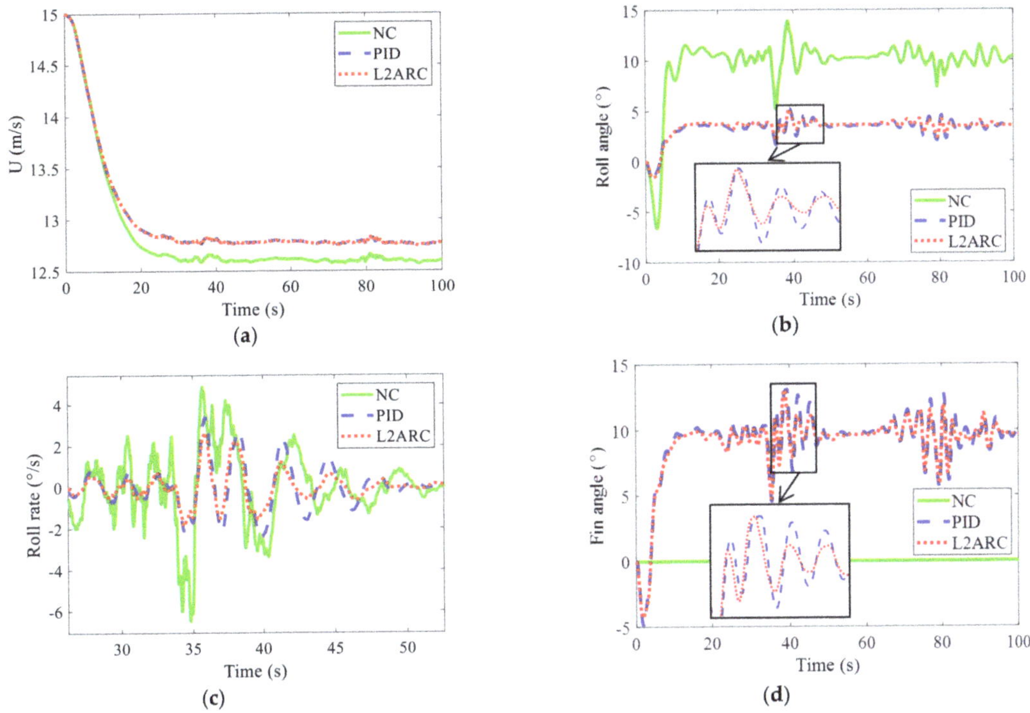

Figure 5. Simulation results of L2-gain adaptive robust control under wave disturbance. (**a**) Ship speed; (**b**) roll angle; (**c**) roll rate; (**d**) fin angle.

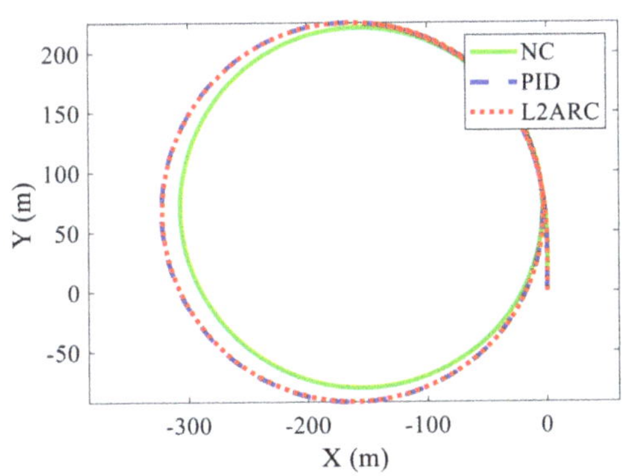

Figure 6. Trajectory of L2-gain adaptive robust control under wave disturbance.

It can be seen from Figure 5 that both the designed L2-gain based adaptive robust controller (L2ARC) and a well-tuned PID controller can effectively reduce the heel and roll motion during ship turns. It can be seen from Figure 5a that the sailing speed of the ship under the action of the controller is slightly increased than that without control, from 12.6 m/s to 12.8 m/s. From Figure 5b,c, it can be seen that the roll and heel reduction control effect of the designed L2ARC is slightly better than a well-tuned PID controller, which shows the superiority of the proposed control strategy. The steady outward heel angle with L2ARC is reduced from 10.19° to 3.01°, and the heel reduction effect can be easily calculated to be about 70.5%. The mean square value of the roll angle during the steady turning phase is reduced from 1.26° to 0.52°, and the roll reduction effect can be easily calculated to be 58.7%. As the encounter angle changes continuously with the progress of the turning, so that the amplitude of the roll motion of the ship also changes periodically. The ship roll is more severe when the encounter angle is 90° or 270°, while other angles are gentler. The control effect of the two controllers is almost the same when the roll is small, but the control effect of L2ARC is better than that of the PID when the roll is large, which is more obvious in Figure 5c. It can be seen from Figure 5d that the fin angle under the action of L2ARC is significant smaller than that under the action of PID. The movement of the fins in the steady turning phase is reduced by approximately 10%, which means the reduction of energy consumption and the effectiveness of the designed L2ARC. The overall effect of the designed L2-gain based adaptive controller is slightly better than well-tuned PID control. In addition, due to the opposite effect of the fin stabilizer on the steady outward heel and turning diameter, the steady turning diameter increases by about 15 m compared with the situation without control, as shown in Figure 6. The simulation results show that the designed controller has good robustness against uncertainties caused by a speed reduction of turning and internal and external disturbance.

6. Conclusions

In this paper, the use of fin stabilizers to reduce the heel and roll during ship turns under wave disturbance is investigated. The 4-DOF nonlinear mathematical model of a multipurpose naval vessel with forces and moments caused by hydrodynamics, waves, fin stabilizers, rudders and propellers is established. Based on the analysis of ship turning motion with different fin angles in calm water, it is found that the fin stabilizer can be used to reduce the heel and roll motion during ship turns, and it has the opposite effect on the steady outward heel and steady turning diameter. Through the analysis, the model of the turning heel/roll reduction system, which is convenient for controller design is obtained. For this disturbed uncertain nonlinear system, the L2-gain based adaptive robust controller is adopted to realize the control of fin stabilizers to reduce the heel and roll during ship turns. Considering the influence of ship speed on the maximum fin angle, fin angle speed adjustment is added in the simulation to get close to the real situation. The simulation results show that both the design L2-based adaptive robust controller and the well-tuned PID controller can effectively reduce the heel and roll motion during ship turns, and the overall effect of the designed controller is slightly better than the well-tuned PID controller.

In the future, we will make further research on the current basis. If possible, we will conduct ship model and even real ship experiments to verify the effectiveness of our proposed method.

Author Contributions: Methodology, Z.S.; writing—original draft preparation, Z.S. and Z.P.; funding acquisition, Z.S. and Z.P.; writing—review and editing, Z.S. and Z.P. All authors have read and agreed to the published version of the manuscript.

Funding: This work is supported by the Doctorial Innovation Funds of Tangshan University under Grant 1402001, the Fundamental Research Funds for the Central Universities under grant HEUCFM170404 and Harbin Science and Technology Innovation Talent Research Special Fund under grant 2017RC2017XK009006.

Data Availability Statement: All the data are available in the article.

Conflicts of Interest: The authors declare no conflict of interest.

References

1. Perez, T. *Ship Motion Control: Course Keeping and Roll Stabilisation Using Rudder and Fins*; Springer: London, UK, 2005.
2. Jin, H.; Yao, X. *Ship Control Principle*, 2nd ed.; Harbin Engineering University: Harbin, China, 2013.
3. Zhang, S.; Liang, L.; Sun, M.; Zhao, P. Fluid motion and stabilization effect prediction of anti-rolling tank coupled ship rolling. *Chin. J. Comput. Mech.* **2016**, *33*, 252–256.
4. Lewis, E. *Principles of Naval Architecture*, 2nd ed.; SNAME: New Jersey, NJ, USA, 1989; Volume 3.

5. Liang, L.; Zhao, P.; Zhang, S.; Yuan, J.; Wen, Y. Simulation and analysis of Magnus rotating roll stabilizer at low speed. *Ocean Eng.* **2017**, *142*, 491–500. [CrossRef]
6. Perez, T.; Blanke, M. Ship roll damping control. *Annu. Rev. Control* **2012**, *36*, 129–147. [CrossRef]
7. Fang, M.-C.; Luo, J.-H. On the track keeping and roll reduction of the ship in random waves using different sliding mode controllers. *Ocean Eng.* **2007**, *34*, 479–488. [CrossRef]
8. Liu, Z.; Jin, H.; Grimble, M.J.; Katebi, R. Ship forward speed loss minimization using nonlinear course keeping and roll motion controllers. *Ocean Eng.* **2016**, *113*, 201–207. [CrossRef]
9. Liang, L.; Wen, Y. Rudder roll stabilization with disturbance compensation model predictive control. *J. Mar. Sci. Technol.* **2019**, *24*, 249–259. [CrossRef]
10. Zhao, P.; Liang, L.; Zhang, S.; Ji, M.; Yuan, J. Simulation analysis of rudder roll stabilization during ship turning motion. *Ocean Eng.* **2019**, *189*, 106322. [CrossRef]
11. Jiguang, S.; Zhao, P.; Liang, L.; Ji, M. Force modeling of zero/low-velocity fin stabilizer and hydrofoil profile optimization. *Ocean Eng.* **2020**, *213*, 107635. [CrossRef]
12. Martin, G. Fin Stabilizers as Maneuver Control Surfaces. Master's Thesis, Naval Postgratude School, Monterey, CA, USA, 2003.
13. Kummer, S.; Hardier, G.; Lambert, C. Heel Compensation for the Charles de Gaulle Aircraft Carrier: Principles and Control Structure. Presented at the RTO AVT Symposium on Fluid Dynamics Problems of Vehicles Operating near or in the Air-Sea Interface, Amsterdam, The Netherlands, 5–8 October 1998.
14. Wang, D. *Research on Effects of Ship Turning Performance under Rudder/Fin Control*; Harbin Engineering University: Harbin, China, 2013.
15. Zhang, S. *Characteristics Research and Experimental Verification of Controlled Passive Tank*; Harbin Engineering University: Harbin, China, 2013.
16. Liang, L.; Liu, H.; Zhang, S. Research of rudder parameters optimization based on ship optimal turning diameter. *Ship Eng.* **2013**, *35*, 61–64.
17. Liu, W. *Research and Simulation of Ship Roll Control in Turning Motion*; Harbin Engineering University: Harbin, China, 2017.
18. Liang, L.; Zhao, P.; Zhang, S.; Ji, M. Simulation analysis of fin stabilizer on ship roll control during turning motion. *Ocean Eng.* **2018**, *164*, 733–748.
19. Hinostroza, M.; Luo, W.; Soares, C. Robust fin control for ship roll stabilization based on L2-gain design. *Ocean Eng.* **2015**, *94*, 126–131. [CrossRef]
20. Crossland, P. The effect of roll-stabilisation controllers on warship operational performance. *Control. Eng. Pract.* **2003**, *11*, 423–431. [CrossRef]
21. Zhang, X. Robust Control for Nonlinear Systems and Its Application to the Lift Feedback Fin Stabilizer. Ph.D. Thesis, Harbin Engineering University, Harbin, China, 2002.
22. Wang, X.; Zhang, X. Nonlinear robust control for rudder/fin joint system with nonlinear course keeping. *Syst. Eng. Electron.* **2008**, *8*, 1549–1552.
23. Li, Z.; Yang, C.; Ding, N.; Bogdan, S.; Ge, T. Robust adaptive motion control for underwater remotely operated vehicles with velocity constraints. *Int. J. Control Autom. Syst.* **2012**, *10*, 421–429. [CrossRef]
24. Kahveci, N.E.; Ioannou, P.A. Adaptive steering control for uncertain ship dynamics and stability analysis. *Automatica* **2013**, *49*, 685–697. [CrossRef]
25. Zhang, G.; Zhang, X. Concise robust adaptive pah-following control of underactuated ships using DSC and MLP. *IEEE J. Ocean. Eng.* **2014**, *39*, 685–694. [CrossRef]
26. Demirel, H.; Alarçin, F. Lmi—based H2 and H state—feedback controller design for fin stabilizer of nonlinear roll motion of a fishing boat. *Brodogr. Teor. Praksa Brodogr. Pomor. Teh.* **2016**, *67*, 91–107. [CrossRef]
27. Zwierzewice, Z. The design of ship autopilot via robust adaptive feedback linearization. In Proceedings of the 21st International Conference on Methods and Models in Automation and Robotics, Miedzyzdroje, Poland, 29 August–1 September 2016; pp. 665–670.
28. Sun, Z.; Zhang, G.; Yi, B.; Zhang, W. Practical proportional integral sliding mode control for underactuated surface ships in the fields of marine practice. *Ocean Eng.* **2017**, *142*, 217–223. [CrossRef]
29. Abkowitz, M. *Lectures on Ship Hydrodynamics-Steering and Manoeuvrability*; Technical Report; Hydro- and Aerodynamics Laboratory: Lyngby, Denmark, 1964.
30. Söder, C.-J.; Rosén, A.; Ovegård, E.; Kuttenkeuler, J.; Huss, M. Parametric roll mitigation using rudder control. *J. Mar. Sci. Technol.* **2013**, *18*, 395–403. [CrossRef]
31. Jia, L.; Yang, Y. *Mathematical Model of Ship Motion*; Dalian Maritime University Press: Dalian, China, 1999.
32. Jin, H.; Wang, K.; Ji, M. *Application of Intelligent Technique to Fin Stabilizers*; National Defense Industry Press: Beijing, China, 2003.
33. Chen, M.; Jiang, C.; Wu, Q.; Cao, B. Adaptive robust L2-gain control for a class of uncertain nonlinear systems. *J. Nanjing Univ. Aeronaut. Astronaut.* **2003**, *35*, 351–355.
34. Yan, D.; He, Y.; Wu, Q. Adaptive robust control with L2-gain for a class of uncertain nonlinear systems. *J. Chang'an Univ. (Nat. Sci. Ed.)* **2006**, *26*, 102–105.

Journal of
*Marine Science
and Engineering*

MDPI

Article

Optimal Damping Concept Implementation for Marine Vessels' Tracking Control

Evgeny I. Veremey

Computer Applications and Systems Department, Saint-Petersburg State University, Universitetskii Prospekt 35, Petergof, 198504 Saint Petersburg, Russia; e_veremey@mail.ru

Abstract: This work presents the results of studies related to the design of stabilizing feedback connections for marine vessels moving along initially given trajectories. As is known, in mathematical formalization, this question leads to a problem of tracking control synthesis for nonlinear and non-autonomous plants. To provide desirable stability and performance features of the closed-loop systems to be synthesized, it is appropriate to use an optimization approach. Unlike the known synthesis methods, which are usually used within the framework of this approach, it is proposed to implement the optimal damping concept first developed by V.I. Zubov in the early 60s of the last century. Modern interpretation of this concept allows constructing numerically effective procedures of control law synthesis taking into account its applicability in a real-time regime. Central attention is focused on the questions connected with practical adaptation of the optimal damping methods for marine control systems. The operability and effectiveness of the proposed approach are illustrated by a practical example of tracking control design.

Keywords: marine vessel; tracking controller; stability; functional; optimal damping

Citation: Veremey, E.I. Optimal Damping Concept Implementation for Marine Vessels' Tracking Control. *J. Mar. Sci. Eng.* **2021**, *9*, 45. https://doi.org/10.3390/jmse9010045

Received: 26 November 2020
Accepted: 29 December 2020
Published: 4 January 2021

Publisher's Note: MDPI stays neutral with regard to jurisdictional claims in published maps and institutional affiliations.

1. Introduction

The nonlinear tracking control of modern marine vessels is one of the most practically significant and theoretically considerable problems in the area of automatic control analysis and design. In particular, tracking control systems are widely used in different branches such as hydrography, inspection of marine constructions, wreck investigation, underwater cable laying, and so on [1,2]. Central theoretical and practical background of tracking control for various moving plants is presented in [3–5] and other fundamental works.

Various issues associated with the design of tracking controllers for marine surface vessels have already been extensively researched and presented in numerous publications (for example, [1,2,6–16]). To evaluate the state of the art in marine tracking control, let us address some modern works presenting this direction of research.

Currently, it is possible to use various ideas to design nonlinear tracking control laws that are reflected in numerous publications, for example, [7,8,13–16]. However, let us note that the mentioned works are not directly oriented to the application of the optimization technique. This makes it difficult to provide the desired dynamic features of the closed-loop connections. Now, it seems to be quite evident that the most effective analytical and numerical tool for feedback connections design is the optimization approach. Several aspects of nonlinear tracking control optimization technique are presented in multitudinous scientific publications, including such popular monographs as [4,5,17–21]. As for the simplest stabilization problem, the autopilots with multipurpose structures of optimized control laws are discussed in detail in [22–26].

The sliding mode control technique for marine control applications is discussed in [1,2,6,8]. This direction seems to be quite constructive, but poorly applicable, since it leads to intensive wear of the actuators.

As for the model predictive control (MPC) approach [9], its most significant disadvantage is the large dimension of the minimization problem that is solved at each step of the control process.

Notably, the complexity of this problem is vast because of the many dynamic requirements, restrictions, and conditions that must be satisfied by the chosen control actions.

It should be noted that many scientific works devoted to the tracking control for marine vessels use linear time invariant models of their motion. However, such models are not quite adequate for the problems of deep maneuvering control in angle and positional dynamic variables. Respectively, one of the most important practical difficulties requiring consideration in the design process is the account of nonlinearity and non-autonomy of the control plant model. In most cases, this problem is a source of dynamic instability and poor performance for various systems that were designed based only on linear approximations.

As for the aforementioned optimization approach, its advantages are determined by the flexibility and convenience of modern optimization methods with respect to the relevant practical demands for control design implementation. Certain analytical and numerical methods are used now to compute the optimal controllers for nonlinear and non-autonomous systems subject to various given performance indices. Nevertheless, there is no saying that the optimization approach is recognized overall as a universal instrument to be put into practice for marine tracking controllers design. This can be explained by the presence of some disadvantages connected with computational troubles. Therefore, there exists a vital necessity to develop persistently analytical and numerical methods of control laws design based on optimization ideology adapting to the specific problems for various marine applications.

At present time, numerous approaches are used for a practical solution of these problems [1–12,17,18]. Usually, they are based on Pontryagin's maximum principle, on Bellman's dynamic programming principle, on finite-dimensional approximation in the range of model predictive control (MPC) technique, etc. Unfortunately, all these approaches are connected with the huge extent of calculations that essentially impedes their implementation both for laboratory design activity and real-time regimes of control.

The existence of numerical difficulties motivates us to use other approaches that allow avoiding the aforementioned shortcomings. This work focuses on a different concept that can be applied to design tracking controllers using the theory of optimal damping (OD). This theory, which was first proposed and developed by V.I. Zubov in his works [19–21], provides effective analytical and numerical methods for control calculations with essentially reduced computational consumptions with respect to classical techniques. We believe that this theory was ahead of its time and was undeservedly underutilized for practical control problem solving. This work is one of the attempts to overcome this omission, taking into account the impressive development of modern computer technologies.

In this article, special attention is paid to the control of marine vessels in terms of the forward speed and heading angle. We are considering the regime of the acceleration in order to achieve the specified forward speed with one-time turn along the heading. To achieve desirable stability and performance features of the reference motion, the correspondent tracking controllers were designed based on the OD technique.

The main contribution of this paper is determined by the following statements. First, we propose to use the OD concept to design tracking controllers for marine vessel speed and heading. This has not been the case before. Second, we discuss a new methodology for selecting the functional to be damped, taking into account the desirable features of the closed-loop system in the range of the optimization technique. Hereby, the choice of this functional as the basis is argued by the guarantee asymptotic stability and the desired quality of control processes. Third, we point to the possibility of applying the OD approach to a wide class of nonlinearities in the mathematical model of the vessel. It is noted that this approach can be implemented in real-time regime of a ship's motion. The practical applicability and effectiveness of the proposed technique is illustrated by a controller design for a transport marine ship.

The novelty of the proposed approach with respect to other works lies in the universality and flexibility of proposed nonlinear non-autonomous control laws based on OD computational procedure, which can be implemented in a real-time regime of functioning for marine control plants.

In general, the present study is an extension of the multipurpose approach proposed in [22–27] and developed in [28] with respect to the marine autopilot control laws with the novel structure, taking into account actuators' time delays.

This article is organized as follows. In Section 2, the optimal damping concept for control law synthesis for nonlinear non-autonomous systems is discussed, taking into account certain specific stability and performance requirements for marine control applications. The known background is presented, and the novel ways are proposed to provide tracing controllers synthesis. Section 3.1 is devoted to the OD synthesis problem statement for the forward speed tracking controller and for the tracking autopilot. Central attention is paid to the presentation of mathematical models of the control plant and dynamic requirements for the quality of the closed-loop connection. Section 3.2 presents an exhaustive novel solution for the mentioned synthesis problem based on the optimal damping concept. In Section 3.3, a practical example of tracking controller synthesis is presented to illustrate the applicability and effectiveness of the proposed approach. Finally, Section 4 concludes the article by discussing the overall results of the investigation and indicates how these results can be further developed.

2. Materials and Methods

As mentioned above, the essence of this paper involves developing an optimal damping technique of tracking control law synthesis for marine vessels with nonlinear and non-autonomous models. In this section, let us first consider the background and some essential features of the OD approach that define the methodological basis of the study.

First of all, let us introduce a commonly used nonlinear robot-like model of the control plant, which represents marine vessel motion for various regimes of its operation [1,2,6]:

$$\begin{aligned} &\mathbf{M}\dot{\mathbf{v}} + \mathbf{C}(\mathbf{v})\mathbf{v} + \mathbf{D}(\mathbf{v})\mathbf{v} + \mathbf{g}(\boldsymbol{\eta}) = \mathbf{G}_u\boldsymbol{\tau} + \mathbf{d}, \\ &\dot{\boldsymbol{\eta}} = \mathbf{J}(\boldsymbol{\eta})\mathbf{v}, \end{aligned} \tag{1}$$

where vector $\mathbf{v} \in R^n$ presents velocities defined in a plant-fixed frame and vector $\boldsymbol{\eta} \in R^n$ contains position dynamical parameters (displacements and angles) in an Earth-fixed frame. External disturbances and controls are presented by the vectors $\mathbf{d} \in R^n$ and $\boldsymbol{\tau} \in R^m$, respectively. Let us accept that the inertia matrix is positive definite: $\mathbf{M} = \mathbf{M}^T > 0$, the matrix of Coriolis-centripetal terms is skew-symmetrical: $\mathbf{C}(\mathbf{v}) = -\mathbf{C}^T(\mathbf{v})$, and the damping matrix $\mathbf{D}(\mathbf{v}) > 0$ is positive definite but non-symmetrical. Vector $\mathbf{g}(\boldsymbol{\eta})$ represents gravitational and buoyancy forces and moments, $\mathbf{J}(\boldsymbol{\eta})$ is the matrix of rotations, and the matrix $\mathbf{G}_u$ with the constant components reflects controls allocation.

Let us provide a transformation of the body-fixed frame representation (1) to the Earth-fixed one with respect to the vector $\boldsymbol{\eta}$. Following [1], this can be done using the following notations:

$$\begin{aligned} &\mathbf{M}_\eta(\boldsymbol{\eta}) := \mathbf{J}^{-T}(\boldsymbol{\eta})\mathbf{M}\mathbf{J}^{-1}(\boldsymbol{\eta}), \\ &\mathbf{C}_\eta(\mathbf{v},\boldsymbol{\eta}) := \mathbf{J}^{-T}(\boldsymbol{\eta})\left[\mathbf{C}(\mathbf{v}) - \mathbf{M}\mathbf{J}^{-1}(\boldsymbol{\eta})\dot{\mathbf{J}}(\boldsymbol{\eta})\right]\mathbf{J}^{-1}(\boldsymbol{\eta}), \\ &\mathbf{D}_\eta(\mathbf{v},\boldsymbol{\eta}) := \mathbf{J}^{-T}(\boldsymbol{\eta})\mathbf{D}(\mathbf{v})\mathbf{J}^{-1}(\boldsymbol{\eta}),\ \ \mathbf{g}_\eta(\boldsymbol{\eta}) := \mathbf{J}^{-T}(\boldsymbol{\eta})\mathbf{g}(\boldsymbol{\eta}), \\ &\boldsymbol{\tau}_\eta := \mathbf{J}^{-T}(\boldsymbol{\eta})\mathbf{G}_u\boldsymbol{\tau},\ \ \mathbf{d}_\eta = \mathbf{J}^{-T}(\boldsymbol{\eta})\mathbf{d}. \end{aligned} \tag{2}$$

In accordance with (2), initial model (1) of the plant takes the form:

$$\ddot{\boldsymbol{\eta}} = -\mathbf{M}_\eta^{-1}(\boldsymbol{\eta})\Big((\mathbf{C}_\eta(\mathbf{v},\boldsymbol{\eta}) + \mathbf{D}_\eta(\mathbf{v},\boldsymbol{\eta}))\dot{\boldsymbol{\eta}} + \mathbf{g}_\eta(\boldsymbol{\eta}) - \boldsymbol{\tau}_\eta - \mathbf{d}_\eta\Big),\ \ \mathbf{v} = \mathbf{J}^{-1}(\boldsymbol{\eta})\dot{\boldsymbol{\eta}}. \tag{3}$$

The essence of the tracking control problem is to provide given desirable motion $\boldsymbol{\eta} = \boldsymbol{\eta}_d(t)$ of the vessel, using the following state feedback

$$\boldsymbol{\tau} = \boldsymbol{\tau}(\boldsymbol{\eta}, \boldsymbol{\nu}, \boldsymbol{\eta}_d(t)), \tag{4}$$

which is a nonlinear non-autonomous tracking controller.

Within mathematical formalization, controller (4) must be implemented to provide the zero equilibrium with respect to the tracking error $\mathbf{e}(t) := \boldsymbol{\eta}(t) - \boldsymbol{\eta}_d(t)$ for the closed-loop system (3), (4), where $\mathbf{d}(t) \equiv 0$. Naturally, this equilibrium point must be asymptotically stable to guarantee that $\mathbf{e}(t) \to 0$ as $t \to \infty$. Let us especially note that the mentioned closed-loop system is nonautonomous, if we have no constant reference motion $\boldsymbol{\eta}_d(t)$. This gives reasons for us to require the uniform asymptotic stability in global (UGAS) or local (UAS) form. An additional requirement is that the controller (4) provides the desired dynamical features for the closed-loop system (3), (4) under the action of an admissible control $\boldsymbol{\tau} \in T_u$.

To set the perform of the controller (4) synthesis, we assume that the vector functions $\boldsymbol{\eta}_d(t)$, $\boldsymbol{\nu}(t) := \mathbf{J}^{-1}(\boldsymbol{\eta}_d(t))\dot{\boldsymbol{\eta}}_d(t)$, and the corresponding $\boldsymbol{\tau}_d(t)$ are given. These functions satisfy Equations (1) or (3), i.e., we have:

$$\begin{aligned} &\mathbf{M}\dot{\boldsymbol{\nu}}_d(t) + \mathbf{C}(\boldsymbol{\nu}_d(t))\boldsymbol{\nu}_d(t) + \mathbf{D}(\boldsymbol{\nu}_d(t))\boldsymbol{\nu}_d(t) + \mathbf{g}(\boldsymbol{\eta}_d(t)) \equiv \mathbf{G}_u\boldsymbol{\tau}_d(t), \\ &\dot{\boldsymbol{\eta}}_d(t) \equiv \mathbf{J}(\boldsymbol{\eta}_d(t))\boldsymbol{\nu}_d(t). \end{aligned} \tag{5}$$

Let introduce the following additional notations:

$$\tilde{\mathbf{x}} := \begin{pmatrix} \tilde{\mathbf{x}}_1 \\ \tilde{\mathbf{x}}_2 \end{pmatrix} = \begin{pmatrix} \boldsymbol{\nu} \\ \boldsymbol{\eta} \end{pmatrix}, \mathbf{f}(\tilde{\mathbf{x}}) := \begin{pmatrix} -\mathbf{M}^{-1}[\mathbf{C}(\boldsymbol{\nu}) + \mathbf{D}(\boldsymbol{\nu})]\boldsymbol{\nu} - \mathbf{M}^{-1}\mathbf{g}(\boldsymbol{\eta}) \\ \mathbf{J}(\boldsymbol{\eta})\boldsymbol{\nu} \end{pmatrix}, \mathbf{B} := \begin{pmatrix} \mathbf{M}^{-1}\mathbf{G}_u \\ 0 \end{pmatrix}$$

which allows us to present Equations (1) and (5) as

$$\dot{\tilde{\mathbf{x}}} = \mathbf{f}(\tilde{\mathbf{x}}) + \mathbf{B}\boldsymbol{\tau}, \dot{\mathbf{x}}_d \equiv \mathbf{f}(\mathbf{x}_d) + \mathbf{B}\boldsymbol{\tau}_d, \tag{6}$$

supposing that $\mathbf{d}(t) \equiv 0$.

Let us also consider deflections

$$\mathbf{x} := \tilde{\mathbf{x}} - \mathbf{x}_d = \begin{pmatrix} \mathbf{x}_1 \\ \mathbf{x}_2 \end{pmatrix} := \begin{pmatrix} \mathbf{e}_\nu \\ \mathbf{e} \end{pmatrix} := \begin{pmatrix} \boldsymbol{\nu} - \boldsymbol{\nu}_d \\ \boldsymbol{\eta} - \boldsymbol{\eta}_d \end{pmatrix}, \ \mathbf{u} := \boldsymbol{\tau} - \boldsymbol{\tau}_d \tag{7}$$

of the vessel dynamical parameters from the desirable motion.

Then, we can present equations of the vessel in the deflections from the desirable motion. Using notations (7) on the base of (6), we obtain

$$\dot{\mathbf{x}} = \boldsymbol{\alpha}(t, \mathbf{x}) + \mathbf{B}\mathbf{u}, \tag{8}$$

where

$$\begin{aligned} \boldsymbol{\alpha}(t,\mathbf{x}) = \boldsymbol{\alpha}(t,\mathbf{e},\mathbf{e}_\nu) := & \begin{pmatrix} -\mathbf{M}^{-1}[\mathbf{C}(\mathbf{e}_\nu + \boldsymbol{\nu}_d(t)) + \mathbf{D}(\mathbf{e}_\nu + \boldsymbol{\nu}_d(t))](\mathbf{e}_\nu + \boldsymbol{\nu}_d(t)) - \mathbf{M}^{-1}\mathbf{g}(\mathbf{e} + \boldsymbol{\eta}_d(t)) \\ \mathbf{J}(\mathbf{e} + \boldsymbol{\eta}_d(t))(\mathbf{e}_\nu + \boldsymbol{\nu}_d(t)) \end{pmatrix} - \\ & \begin{pmatrix} -\mathbf{M}^{-1}[\mathbf{C}(\boldsymbol{\nu}_d(t)) + \mathbf{D}(\boldsymbol{\nu}_d(t))]\boldsymbol{\nu}_d(t) - \mathbf{M}^{-1}\mathbf{g}(\boldsymbol{\eta}_d(t)) \\ \mathbf{J}(\boldsymbol{\eta}_d(t))\boldsymbol{\nu}_d(t) \end{pmatrix}. \end{aligned} \tag{9}$$

It is a matter of simple calculations to check that equation (6) has zero equilibrium position, which must be stabilized by the choice of the controller $\mathbf{u} = \mathbf{u}(t, \mathbf{x})$. If this controller is found, then the tracking feedback (4) can be presented as

$$\boldsymbol{\tau} = \boldsymbol{\tau}(t, \boldsymbol{\nu}, \boldsymbol{\eta}) = \boldsymbol{\tau}(\boldsymbol{\eta}, \boldsymbol{\nu}, \boldsymbol{\eta}_d) = \mathbf{u}(t, \mathbf{x}) + \boldsymbol{\tau}_d(t), \ \mathbf{x} := \begin{pmatrix} \boldsymbol{\nu} - \boldsymbol{\nu}_d(t) \\ \boldsymbol{\eta} - \boldsymbol{\eta}_d(t) \end{pmatrix}. \tag{10}$$

As for the desirable dynamic features of the closed-loop connection, the most widely used formalized approach is based on minimizing the following integral functional:

$$J = J(t, \mathbf{x}, \mathbf{u}) = \int_{t_0}^{\infty} f(\boldsymbol{\tau}, \mathbf{x}, \mathbf{u})\, d\boldsymbol{\tau},$$

that determines the quality of control processes for a closed-loop system. Here, subintegral function f is positive definite, i.e., $f(t, \mathbf{x}, \mathbf{u}) \geq 0 \ \ \forall t \geq t_0, \ \forall \mathbf{x}, \ \forall \mathbf{u}, \ f = 0 \Leftrightarrow \mathbf{x} = 0, \ \mathbf{u} = 0.$ However, as is well known, there are certain difficulties in directly implementing this approach, which are determined by the huge extent of calculations that essentially impedes their practical implementation.

We propose to overcome the mentioned obstacles using a novel technique based on the OD concept, connected to the OD problem for the synthesis of the control $\mathbf{u}$. To state this problem, firstly, let us introduce the functional to be damped as follows:

$$L = L(t, \mathbf{x}, \mathbf{u}) = V(t, \mathbf{x}) + \int_{t_0}^{t} f_0(\boldsymbol{\tau}, \mathbf{x}, \mathbf{u})\, d\boldsymbol{\tau}, \tag{11}$$

where $V = V(t, \mathbf{x})$ is a Lyapunov function candidate, and f_0 is a positively defined function. Let us additionally accept that the function V satisfies the following conditions:

$$\alpha_1(\|\mathbf{x}\|) \leq V(t, \mathbf{x}) \leq \alpha_2(\|\mathbf{x}\|) \forall \mathbf{x} \in B_r \subset E^n, \ \forall t \geq t_0, \tag{12}$$

where $\alpha_1, \alpha_2 \in K$ are comparison functions [4].

The essence of the OD approach consists of the control generation as a function from the current values of variables $t, \mathbf{x}$ in the form

$$\mathbf{u}_0 = \mathbf{u}_0(t, \mathbf{x}) = \underset{\mathbf{u} \in U}{\operatorname{argmin}} W(t, \mathbf{x}, \mathbf{u}) \tag{13}$$

where $U \subset E^m$ is the metric compact set, and W is a rate of the functional L change along the motions of the plant (8):

$$W(t, \mathbf{x}, \mathbf{u}) := \left.\frac{dL}{dt}\right|_{(8)} = \left.\frac{dV}{dt}\right|_{(8)} + f_0(t, \mathbf{x}, \mathbf{u}) = \frac{\partial V}{\partial t}(t, \mathbf{x}) + \frac{\partial V}{\partial \mathbf{x}}(t, \mathbf{x})\, \boldsymbol{\alpha}(t, \mathbf{x}) + \frac{\partial V}{\partial \mathbf{x}}(t, \mathbf{x})\, \mathbf{B}\mathbf{u} + f_0(t, \mathbf{x}, \mathbf{u}).$$

In other words, it is necessary to find OD controller (13), using an admissible set $U \subset E^m$ such that $\forall \mathbf{u} \in U : \boldsymbol{\tau}_d(t) + \mathbf{u}(t, \mathbf{x}) \in T_u, \ \forall t \geq 0.$

We can specify three possible ways to solve this optimization problem:

(a) The first way is based on the direct numerical calculation of the vectors $\mathbf{u} = \mathbf{u}_0(t, \mathbf{x})$ providing the pointwise minimization of the function W by the choice of $\mathbf{u}$ for current points $(t, \mathbf{x})$. Let us especially notice that this variant is universal in nature and can be applied to generate a control signal for real-time regime of motion.

(b) The second way involves the possibility of an analytical solution to the problem (13). Naturally, this is the most preferred way, but such a situation is quite rare, although an example of its practical application will be given below.

(c) The third way reduces the problem (13) to a numerical solution of a nonlinear system of finite equations. In fact, if we have $\mathbf{u}_0(t, \mathbf{x}) \in \operatorname{int} U \ \forall t \geq 0, \ \forall \mathbf{x} \in B_r$, then with necessity we obtain

$$\left.\frac{dW}{d\mathbf{u}}\right|_{\mathbf{u} = \mathbf{u}_0(t, \mathbf{x})} = 0 \ \Rightarrow \ \left[\frac{\partial V}{\partial \mathbf{x}}(t, \mathbf{x})\, \mathbf{B} + \frac{\partial F}{\partial \mathbf{u}}(t, \mathbf{x}, \mathbf{u})\right]_{\mathbf{u} = \mathbf{u}_0(t, \mathbf{x})} = 0. \tag{14}$$

Using the necessary condition (14), one can solve the following nonlinear system

$$\mathbf{a}(t, \mathbf{x}, \mathbf{u}) = \mathbf{b}(t, \mathbf{x})$$

for any point $(t, \mathbf{x})$ with respect to the vector $\mathbf{u}$, where $\mathbf{a}(t, \mathbf{x}, \mathbf{u}) = \text{col}[a_i(t, \mathbf{x})]$, $\mathbf{b}(t, \mathbf{x}) = \text{col}[b_i(t, \mathbf{x})]$,

$$a_i(t, \mathbf{x}, \mathbf{u}) := \left[\frac{\partial F}{\partial \mathbf{u}}(t, \mathbf{x}, \mathbf{u})\right]_i, \quad b_i(t, \mathbf{x}) := -\left[\frac{\partial V}{\partial \mathbf{x}}(t, \mathbf{x})\, \mathbf{B}\right]_i, \quad i = \overline{1, n}.$$

Based on [4,5], it can be shown that if the function $V = V(t, \mathbf{x})$ is such that $W(t, \mathbf{x}, \mathbf{u}_0(t, \mathbf{x})) \leq -\alpha_3(\|\mathbf{x}\|)$ $\forall \mathbf{x} \in B_r$, $\forall t \geq t_0$, where $\alpha_3 \in K$, then the function V is control Lyapunov function (CLF) for the plant (8), and the zero equilibrium for the closed-loop system (8), (13) is UAS.

3. Results

This section is devoted to a practical implementation of the proposed OD approach to nonlinear tracking controllers design for marine vessels. Particular attention is given to the forward speed tracking control law with initially given reference signal.

3.1. Tracking Control Problem for Marine Vessels

To consider the problems of tracking control for marine applications, let us accept the following widely used [1,2] nonlinear dynamical model of a marine vessel:

$$\begin{array}{l} \frac{d\overline{V}_x}{dt} = T_v(\overline{V}_x, \overline{\xi}, \overline{u}_1) + T_h(\overline{V}_x, \overline{\xi}), \\ \frac{d\overline{V}_y}{dt} = h_2(\overline{V}_x, \overline{\xi}), \ \frac{d\overline{\omega}}{dt} = h_3(\overline{V}_x, \overline{\xi}), \\ \frac{d\overline{\varphi}}{dt} = \overline{\omega}, \ \frac{d\overline{\delta}}{dt} = \overline{u}_2. \end{array} \tag{15}$$

Here, the following notations are used: $\overline{V}_x$, $\overline{V}_y$, and $\overline{\omega}$ are the projections of the velocity vectors on the axes of a vessel-fixed frame $Oxyz$ (Figure 1); $\overline{\xi} := \begin{pmatrix} \overline{V}_y & \overline{\omega} & \overline{\varphi} & \overline{\delta} \end{pmatrix}^T$ is the auxiliary vector of dynamical parameters; $\overline{\varphi}$ is the heading angle, and $\overline{\delta}$ is the vertical rudder deflection. The functions T_v and T_h represent hydrodynamical forces, which are produced by the ship's engine and the water resistance correspondingly.

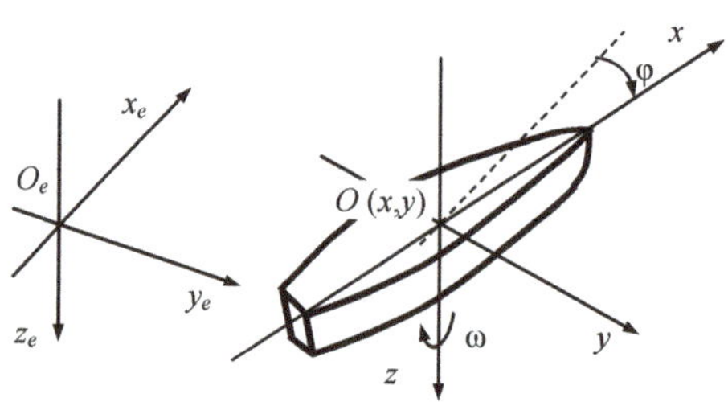

Figure 1. Earth-fixed and vessel-fixed coordinate frames.

The control signals $\overline{u}_1$ and $\overline{u}_2$ must be composed by the automatic control system to be designed. The first of them determines a reference surge speed of the vessel, and the second one presents a reference speed of the rudders' deflections.

Assuming that the number of the screw rotations is proportional to a reference surge speed, the function $T_v(\overline{V}_x, \overline{\xi}, \overline{u}_1)$ determining a trust force of the screw can be presented in the form

$$T_v(\overline{V}_x, \overline{\xi}, \overline{u}_1) \equiv \alpha \overline{u}_1^2 + \beta(\overline{V}_x, \overline{\xi})\overline{u}_1 + \widetilde{\gamma}(\overline{V}_x, \overline{\xi}), \tag{16}$$

We accept here that $\alpha \equiv const$, and that the variables β and $\widetilde{\gamma}$ are treated as known functions of the dynamical parameters $\overline{V}_x, \overline{\xi}$.

Let us introduce new vector variables $\mathbf{h}(\overline{V}_x, \overline{\xi}) := \begin{pmatrix} h_2(\overline{V}_x, \overline{\xi}) & h_1(\overline{V}_x, \overline{\xi}) & \overline{\xi}_3 & 0 \end{pmatrix}^T$, $\mathbf{b} = \begin{pmatrix} 0 & 0 & 0 & 1 \end{pmatrix}^T$, and, taking into account (16), we can rewrite nonlinear model (1) of the vessel dynamic as follows:

$$\begin{aligned} \frac{d\overline{V}_x}{dt} &= \alpha\overline{u}_1^2 + \beta(\overline{V}_x, \overline{\xi})\overline{u}_1 + \gamma(\overline{V}_x, \overline{\xi}), \\ \frac{d\overline{\xi}}{dt} &= \mathbf{h}(\overline{V}_x, \overline{\xi}) + \mathbf{b}\overline{u}_2. \end{aligned} \tag{17}$$

Now we specify a certain reference motion $V_{x\rho}(t)$, $\xi_\rho(t)$, $u_{1\rho}(t)$, and $u_{2\rho}(t)$ of the plant (17), satisfying the identities

$$\begin{aligned} \frac{dV_{x\rho}}{dt} &\equiv \alpha u_{1\rho}^2 + \beta(V_{x\rho}, \xi_\rho)u_{1\rho} + \gamma(V_{x\rho}, \xi_\rho), \\ \frac{d\xi_\rho}{dt} &\equiv \mathbf{h}(V_{x\rho}, \xi_\rho) + \mathbf{b}u_{2\rho}. \end{aligned} \tag{18}$$

Using systems (17) and (18), we can present equations of vessel dynamic in deflections from the desirable reference motion of the form

$$\begin{aligned} \frac{dV_x}{dt} + \frac{dV_{x\rho}}{dt} &= \alpha(u_1 + u_{1\rho})^2 + \beta(V_x + V_{x\rho}, \xi + \xi_\rho)(u_1 + u_{1\rho}) + \\ &\quad + \gamma(V_x + V_{x\rho}, \xi + \xi_\rho), \\ \frac{d\xi}{dt} + \frac{d\xi_\rho}{dt} &= \mathbf{h}(V_x + V_{x\rho}, \xi + \xi_\rho) + \mathbf{b}(u_2 + u_{2\rho}), \end{aligned} \tag{19}$$

where

$$V_x := \overline{V}_x - V_{x\rho}, \ \xi := \overline{\xi} - \xi_\rho, \ u_1 := \overline{u}_1 - u_{1\rho}, \ u_2 := \overline{u}_2 - u_{2\rho}. \tag{20}$$

One can see that $V_{x\rho} = V_{x\rho}(t)$, $\xi_\rho = \xi_\rho(t)$, $u_{1\rho} = u_{1\rho}(t)$, and $u_{2\rho} = u_{2\rho}(t)$ are known functions of t: using new correspondent notations β_1, γ_1, $\mathbf{h}_1$, we can rewrite (19) as follows:

$$\begin{aligned} \frac{dV_x}{dt} &= \alpha u_1^2 + \beta_1(t, V_x, \xi)u_1 + \gamma_1(t, V_x, \xi), \\ \frac{d\xi}{dt} &= \mathbf{h}_1(t, V_x, \xi) + \mathbf{b}u_2 \end{aligned} \tag{21}$$

stating that the resulting system (21) has a zero-equilibrium position.

Let us especially note that, unlike (17), this system is not only nonlinear, but also non-autonomous. This is due to the explicit introduction of the time-dependent reference signals into the vessel dynamics equations.

The purpose of the control design procedure is to construct the following stabilizing feedback controllers in deviations:

$$u_1 = u_1(t, V_x, \xi), \ u_2 = u_2(t, V_x, \xi), \tag{22}$$

such that the zero equilibrium of the closed-loop connection (21), (22) is asymptotically stable. This means that if the motion of the initial plant (15) takes place under the action of tracking controllers of the form

$$\overline{u}_1 = u_{1\rho}(t) + u_1(t, V_x, \xi), \ \overline{u}_2 = u_{2\rho}(t) + u_2(t, V_x, \xi), \tag{23}$$

starting in some neighborhood of the point $\{V_{x\rho}(0), \ \xi_\rho(0)\}$, then this motion tends to the reference one as t becomes infinite.

As for the performance of control processes, they are usually formalized mathematically using certain functionals, which are given on the motions of the closed-loop system (21), (22). Currently, the commonly used approach to design stabilizing controllers (22) is setting and solving the following optimization problem

$$J = J(u_1, u_2) \to \min_{\{u_1, u_2\} \in \widetilde{U}} \tag{24}$$

based on the integral functional

$$J = J(u_1, u_2) = \int_{t_0}^{\infty} f^*(t, V_x(t), \xi(t), u_1(t), u_2(t))\, dt, \tag{25}$$

where $\widetilde{U}$ is the set of admissible pairs $\{u_1, u_2\}$ and subintegral function f^* is positive definite for all its arguments.

In contrast to the problems (24), (25) with traditional methods of its solving, as mentioned above, we propose to use novel approach based on the OD concept.

Let us especially note that both the solution of the problem (24) and the solution of the OD problem significantly depend on the initially given mathematical model (21) of a marine vessel. Naturally, this is evidence that this model cannot be formed accurately, which raises very important questions about the robust features of the closed-loop system to be designed. This problem is quite solvable for various types of marine vessels: the paper [29] is an example. However, the study of the robust features of tracking OD controllers presents an independent problem to be addressed in future studies.

3.2. Forward Speed OD Tracking Controller Design

In general, the synthesis of the two stabilizing controllers (22) using OD technique can be carried out simultaneously. Nevertheless, one can also apply a combined approach to the feedback design for the considered plant (21) with two control channels. In the range of this approach, the first control is formed based on the OD problem, and the second one can be designed in any other way, providing desirable performance features. However, a joint closed-loop system must have zero equilibrium, and this motion must be asymptotically stable.

Realizing this idea, let us accept the dynamic controller for rudders (marine autopilot) in the following form:

$$\begin{aligned} \tfrac{d\mathbf{z}}{dt} &= \boldsymbol{\gamma}_z(t, \mathbf{z}, V_x, \xi), \\ u_2 &= g_z(t, \mathbf{z}, V_x, \xi), \end{aligned} \tag{26}$$

where $\mathbf{z} \in E^{\nu}$ is the state vector of this controller.

In particular, the feedback (26) may have a multipurpose structure, which is presented in detail in [22,27] for linear time-invariant case. Its implementation taking into account control time-delay is investigated in the paper [28].

The equations of the control plant now take the form

$$\begin{aligned} \tfrac{dV_x}{dt} &= \alpha u_1^2 + \beta_1(t, V_x, \xi)u_1 + \gamma_1(t, V_x, \xi), \\ \tfrac{d\xi}{dt} &= \mathbf{h}_1(t, V_x, \xi) + \mathbf{b}g_z(t, \mathbf{z}, V_x, \xi), \\ \tfrac{d\mathbf{z}}{dt} &= \boldsymbol{\gamma}_z(t, \mathbf{z}, V_x, \xi). \end{aligned} \tag{27}$$

In order to synthesize a feedback for the first control channel, i.e., design the OD forward speed stabilizing controller, let us introduce the functional to be damped as

$$L = L(t, V_x, \xi, \mathbf{z}, u_1) = V(V_x, \xi, \mathbf{z}) + \int_{t_0}^{t} \lambda_1^2 u_1^2\, d\tau. \tag{28}$$

Let us take as a Lyapunov function candidate the following sum of quadratic forms:

$$V = V(V_x, \xi, \mathbf{z}) = \lambda^2 V_x^2 + \xi^T \mathbf{Q} \xi + \mathbf{z}^T \mathbf{Q}_1 \mathbf{z}, \ \ \mathbf{Q} > 0, \ \mathbf{Q}_1 > 0. \tag{29}$$

Based on (28), (29), we can state the correspondent OD problem of the form

$$W(t, V_x, \xi, \mathbf{z}, u_1) \to \min_{u_1 \in U_1} \tag{30}$$

where the rate W of the functional L change is determined as follows:

$$W = W(t, V_x, \boldsymbol{\xi}, \mathbf{z}, u_1) := \left.\tfrac{dL}{dt}\right|_{(27)} = \left.\tfrac{dV}{dt}\right|_{(27)} + \lambda_1^2 u_1^2 = \tfrac{\partial V}{\partial V_x}\tfrac{dV_x}{dt} + \tfrac{\partial V}{\partial \boldsymbol{\xi}}\tfrac{d\boldsymbol{\xi}}{dt} + \tfrac{\partial V}{\partial \mathbf{z}}\tfrac{d\mathbf{z}}{dt} + \lambda_1^2 u_1^2$$
$$= 2\lambda^2 V_x\left[\alpha u_1^2 + \beta_1(t, V_x, \boldsymbol{\xi})u_1 + \gamma_1(t, V_x, \boldsymbol{\xi})\right] + 2\boldsymbol{\xi}^T\mathbf{Q}\tfrac{d\boldsymbol{\xi}}{dt} + 2\mathbf{z}^T\mathbf{Q}_1\tfrac{d\mathbf{z}}{dt} + \lambda_1^2 u_1^2. \quad (31)$$

Let us solve the problem (30), taking into account (31) and assuming that the extreme is achieved at the inner point of the set U_1: with necessity we obtain

$$\frac{dW}{du_1} = 2\lambda^2 V_x(2\alpha u_1 + \beta_1(t, V_x, \boldsymbol{\xi})) + 2\lambda_1^2 u_1 = 0$$

that determines the following controller

$$u_1 = u_{10}(t, V_x, \boldsymbol{\xi}) = -\frac{\lambda^2 V_x \beta_1(t, V_x, \boldsymbol{\xi})}{2\lambda^2 V_x \alpha + \lambda_1^2}. \quad (32)$$

Since, according to formulae (15)–(21), we have

$$\beta_1 = \beta_1(t, V_x, \boldsymbol{\xi}) := 2\alpha u_{1\rho}(t) + \beta(V_x + V_{x\rho}(t), \boldsymbol{\xi} + \boldsymbol{\xi}_\rho(t)),$$
$$\beta(V_x + V_{x\rho}(t), \boldsymbol{\xi} + \boldsymbol{\xi}_\rho(t)) = \beta_0\sqrt{(V_x + V_{x\rho}(t))^2 + (V_y + V_{y\rho}(t))^2}, \ \beta_0 = \text{const},$$

we arrive from (32) at the following OD controller:

$$u_{10}(t, V_x, \boldsymbol{\xi}) = -\frac{\lambda^2 V_x\left(2\alpha u_{1\rho}(t) + \beta_0\sqrt{(V_x + V_{x\rho}(t))^2 + (V_y + V_{y\rho}(t))^2}\right)}{2\lambda^2 V_x \alpha + \lambda_1^2}, \quad (33)$$

where $V_y = \begin{pmatrix} 1 & 0 & 0 & 0 \end{pmatrix}\boldsymbol{\xi}$, $V_{y\rho}(t) = \begin{pmatrix} 1 & 0 & 0 & 0 \end{pmatrix}\boldsymbol{\xi}_\rho(t)$.

It is necessary to note that if the value $u_{10}(t, V_x, \boldsymbol{\xi})$ is out of an inner part of the admissible set U_1, we must use the control signal $u_1 = 0.01P_\rho u_{1\rho}(t) \cdot \text{sign}(u_{10}(t, V_x, \boldsymbol{\xi}))$ instead of (33).

It is necessary to note that practical problems involve situations where the right part of the equation for forward speed has a more complex structure than for the system (17). In general, this equation can be presented as

$$\frac{d\overline{V}_x}{dt} = F_x(\overline{V}_x, \overline{\boldsymbol{\xi}}, \overline{u}_1).$$

which results in the correspondent equation for the system (21) in deflections:

$$\frac{dV_x}{dt} = G_x(t, V_x, \boldsymbol{\xi}, u_1),$$

where $G_x(t, V_x, \boldsymbol{\xi}, u_1) := F_x(V_x + V_{x\rho}(t), \boldsymbol{\xi} + \boldsymbol{\xi}_\rho(t), u_1 + u_{1\rho}(t)) - F_x(V_{x\rho}(t), \boldsymbol{\xi}_\rho(t), u_{1\rho}(t))$. In this case, instead of (31) we have

$$W(t, V_x, \boldsymbol{\xi}, \mathbf{z}, u_1) = 2\lambda^2 V_x G_x(t, V_x, \boldsymbol{\xi}, u_1) + 2\boldsymbol{\xi}^T\mathbf{Q}\frac{d\boldsymbol{\xi}}{dt} + 2\mathbf{z}^T\mathbf{Q}_1\frac{d\mathbf{z}}{dt} + \lambda_1^2 u_1^2.$$

and the analytical search for stationary points of the function W becomes problematic. Nevertheless, it is always possible to consider a question about the numerical solution of OD problem (30) for each aggregate of fixed parameters $t, V_x, \boldsymbol{\xi}$. Moreover, it is always possible to pose a finite dimensional minimization problem

$$u_1 = u_{10}^n(t, V_x, \boldsymbol{\xi}) = \arg\min_{u_1 \in U_{1n}}\left(g_1(t, V_x, \boldsymbol{\xi}, u_1) + \lambda_1^2 u_1^2\right), \quad (34)$$

on the finite net $U_{1n} \subset U_1$ for any compact set U_1. This problem should be solved at the time t for the fixed parameters $t, V_x(t), \boldsymbol{\xi}(t)$. Obviously, with a sufficiently large number of points for the set U_{1n}, the solution of the problem (34) tends to solution for the same problem on the set U_1.

Finally, in accordance with formula (23), we can compose the tracking controller of the form

$$\overline{u}_1 = \overline{u}_{01}(t, \overline{V}_x, \overline{\xi}) := u_{1\rho}(t) + u_{10}(t, V_x, \xi),$$

where the stabilizing part $u_{10}(t, V_x, \xi)$ is determined by equality (33). One can see that the resulting representation for OD forward speed tracking controller is as follows:

$$\overline{u}_1 = \overline{u}_{01}(t, \overline{V}_x, \overline{V}_y) = \frac{\lambda_1^2 u_{1\rho}(t) - \lambda^2(\overline{V}_x - V_{x\rho}(t))\,\beta_0\sqrt{\overline{V}_x{}^2 + \overline{V}_y{}^2}}{2\lambda^2\alpha(\overline{V}_x - V_{x\rho}(t)) + \lambda_1^2}. \tag{35}$$

The current values of the dynamic parameters $\overline{V}_x(t)$, $\overline{V}_y(t)$ must be measured to implement the controller (35).

3.3. Numerical Example of OD Tracking Controller Synthesis

To illustrate a practical implementation of the proposed OD approach, let us consider a practical example of forward speed tracking controller design for the transport ship with a displacement of 3500 tons, a length of 110 m, a width of 14 m, and an immersion of 5 m. As a mathematical model of the plant, let us accept equations (17), which are presented in [1,2] with the given functions $\beta(\overline{V}_x, \overline{\xi})$, $\gamma(\overline{V}_x, \overline{\xi})$, $\mathbf{h}(\overline{V}_x, \overline{\xi})$ and parameter α.

First, we define the reference motion of the ship as a partial solution of system (17) with the initial conditions $\overline{V}_x(0) = 2\ m/s$, $\xi(0) = 0$. Let us accept the reference forward speed control as follows:

$$u_{1\rho}(t) = \begin{cases} 2 + 0.08t, & if\ 0 \le t \le 100s, \\ 10, & if\ t \ge 100s. \end{cases} \tag{36}$$

To determine the reference control for the rudders, let us initially construct a model feedback as a controller

$$u_m = k_1 x_{m1} + k_2 x_{m2} + k_3(x_{m3} - \varphi_z) + k_4 x_{m4} := \mathbf{k}_m \mathbf{x}_m - k_3\varphi_z, \tag{37}$$

fulfilling the command-heading signal φ_z. The basis item $\mathbf{k}_m\mathbf{x}_m$ in (37) stabilizes Linear Time Invariant (LTI) plant

$$\frac{d\mathbf{x}_m}{dt} = \mathbf{A}_m\mathbf{x}_m + \mathbf{b}_m u_m \tag{38}$$

which is a result of the plant (17) linearization in the neighborhood of the origin for a fixed forward speed $\overline{V}_x$. Let us especially notice that the linear model (38) is used only for the synthesis of controller (37) but is not used for simulation and for performance indices computation: controller (37) closes full initial model (17) of the vessel.

Accepting $\overline{V}_x = 10$ m/s , we obtain

$$\mathbf{A}_m = \begin{pmatrix} a_{11} & a_{12} & 0 & b_1 \\ a_{21} & a_{22} & 0 & b_2 \\ 0 & 1 & 0 & 0 \\ 0 & 0 & 0 & 0 \end{pmatrix},\ \mathbf{b}_m = \begin{pmatrix} 0 \\ 0 \\ 0 \\ 1 \end{pmatrix}\quad \begin{matrix} a_{,11} = -0.0936, & a_{12} = -6.34, & b_1 = -0.190, \\ a_{21} = -0.00480, & a_{22} = -0.717, & b_2 = 0.0160. \end{matrix}$$

We design the basis control $u_m = \mathbf{k}_m\mathbf{x}_m$ for plant (38) with aforementioned parameters as the Linear Quadratic Regulator (LQR) optimal controller with respect to the functional

$$J_m = J(u_m) = \int_0^\infty \left(\mathbf{x}_m^T\mathbf{Q}_m\mathbf{x}_m + \lambda_m u_m^2\right) dt, \tag{39}$$

where $\mathbf{Q}_m = \mathrm{diag}(\ 0\quad 1.975\quad 0.0250\quad 0\ )$, $\lambda_m = 60$. Let us especially remark that the choice of the presented parameters for the functional (39) is determined by the initially given requirements to a dynamic quality of the control processes for the nonlinear time varying system. These requirements provide desirable settling time and overshoot for the closed-loop connection.

As a result of computations, we get the constant vector $\mathbf{k}_m = \begin{pmatrix} k_1 & k_2 & k_3 & k_4 \end{pmatrix}$, where

$$k_1 = 0.00234,\ k_2 = -0.0495,\ k_3 = -0.0204,\ k_4 = -0.0497.$$

Let us substitute the reference heading control $u_{1\rho}(t)$ and the rudders feedback control

$$\overline{u}_2 = k_1\overline{\xi}_1 + k_2\overline{\xi}_2 + k_3(\overline{\xi}_3 - \varphi_z) + k_4\overline{\xi}_4,\ \varphi_z = 90^\circ, \tag{40}$$

into Equation (17). After integration, we obtain the functions $u_{2\rho} = u_{2\rho}(t)$, $V_{x\rho} = V_{x\rho}(t)$, and $\xi_\rho = \xi_\rho(t) = (V_{y\rho}(t)\ \omega_\rho(t)\ \varphi_\rho(t)\ \delta_\rho(t))^T$, which determine given reference motion of the vessel. The graphs of aforementioned functions are presented in Figures 2 and 3.

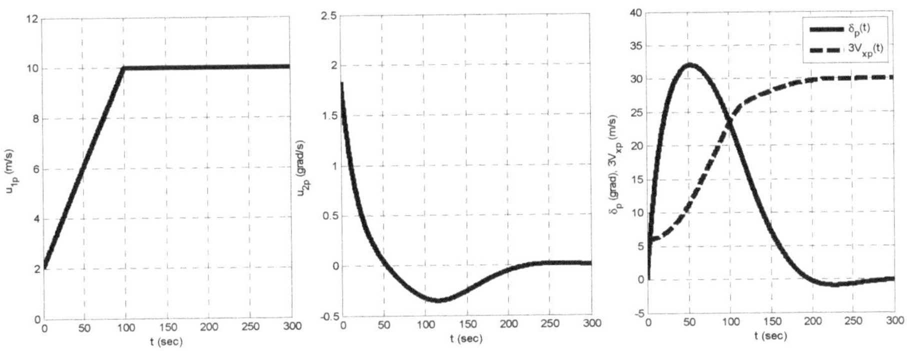

Figure 2. Graphs of the reference functions $u_{1\rho}(t)$, $u_{2\rho}(t)$, $V_{x\rho}(t)$, and $\delta_\rho(t)$.

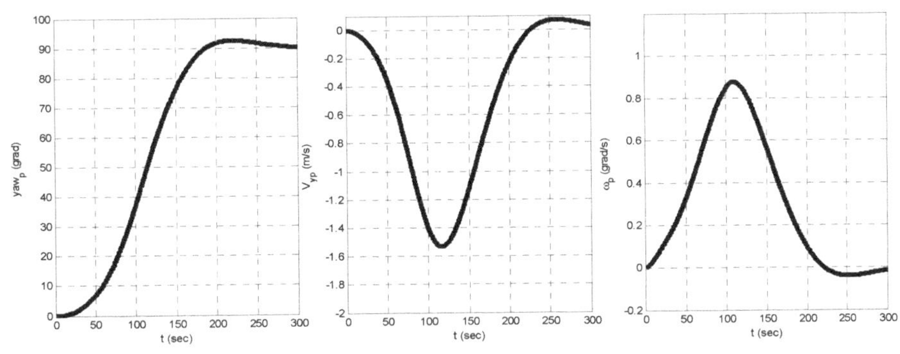

Figure 3. Graphs of the reference functions $\varphi_\rho(t)$, $V_{y\rho}(t)$, and $\omega_\rho(t)$.

To provide simulation process, instead of (23), (26), we accept the following simple feedback control law for a marine autopilot:

$$\overline{u}_2 = k_1\overline{V}_y + k_2\overline{\omega} + k_3(\overline{\varphi} - \varphi_z) + k_4\overline{\delta}, \tag{41}$$

where φ_z is the command-heading signal. This controller, corresponding to (40), can be directly used for actuators of the vertical rudders.

Thus, one can see that, as a result, initial control plant (17) is closed by the tracking controllers (35) and (41). The current values of the dynamic parameters $\overline{V}_x(t)$, $\overline{V}_y(t)$, $\overline{\omega}(t)$,

$\overline{\varphi}(t)$, and $\overline{\delta}(t)$ must be measured by the corresponding sensors for the actual implementation of these controllers.

For simulation of the closed-loop system dynamics, the following parameters values are accepted: $\lambda^2 = 100$, $\lambda_1^2 = 1$, $\alpha = 0.00462$, $\beta_0 = -0.00322$. In addition, let us take into account the restrictions $|\overline{\delta}(t)| \leq d_m = 35°$ and $|\overline{u}_2(t)| \leq u_m = 3\,°/\text{c}$.

To illustrate the practical applicability of the proposed approach, let us simulate the control processes for the closed-loop connection. The aim is to make the proposed approach comparable to other methods. This determined the choice of design parameters and regimes of vessel's motion. These regimes represent the most popular options for movement on quiet water and under the action of sea waves.

The results of simulation are presented in Figures 4–6 as the graphs of corresponding functions, which reflect control signals and the vessel's state variables for the transient process. This process is determined by the aforementioned reference motion, which is realized with the help of the designed tracking controllers. Initial conditions for all variables are zero with the exception of forward speed and heading angle. By these variables, the initial conditions $\overline{V}_x(0) = 4$ m/s and $\overline{\varphi}(0) = -10°$ are accepted to distinguish them from the reference motion.

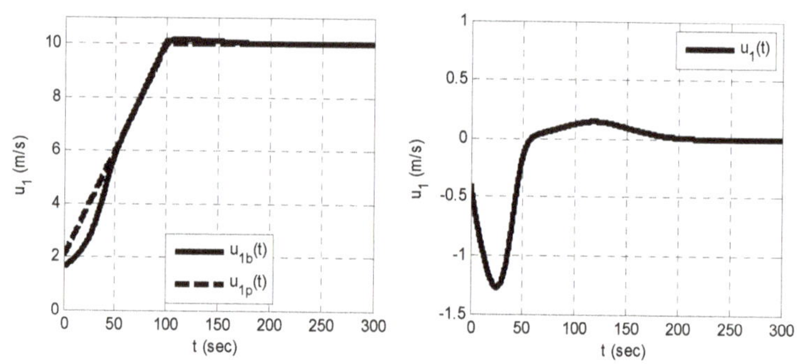

Figure 4. First control signals $\overline{u}_1 = u_{1b}(t), u_{1\rho}(t)$, and $u_1(t) = \overline{u}_1(t) - u_{1\rho}(t)$.

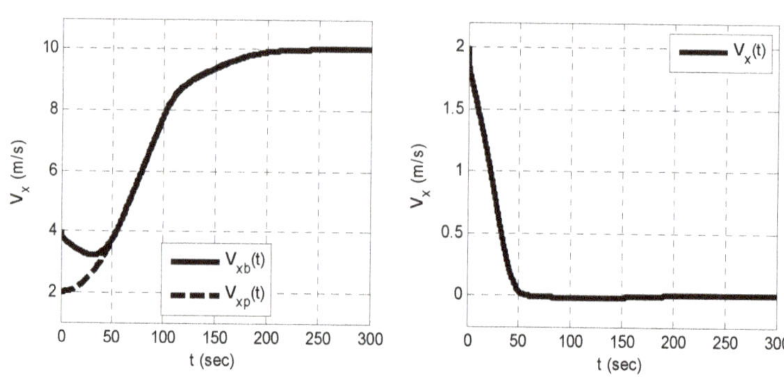

Figure 5. Forward speed $\overline{V}_x = V_{xb}(t), V_{x\rho}(t)$, and $V_x(t) = \overline{V}_x(t) - V_{x\rho}(t)$.

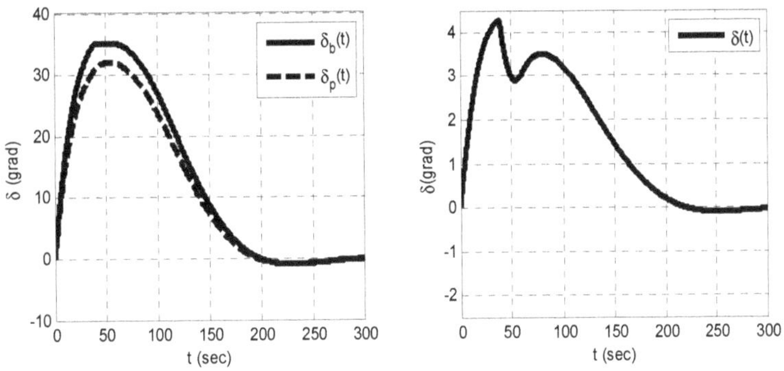

Figure 6. Rudders deflections $\overline{\delta} = \delta_b(t), \delta_\rho(t)$, and $\delta(t) = \overline{\delta}(t) - \delta_\rho(t)$.

Let us note that the dynamical quality of the presented transient process seems to be quite satisfactory. In addition, it is suitable to illustrate the dynamics of the closed-loop connection under the action of sea wave external disturbance. Figure 7 shows the same control process for the forward speed $\overline{V}_x$, taking into account presence the approximate representation of sea waves with an intensity of 5 on the Beaufort scale.

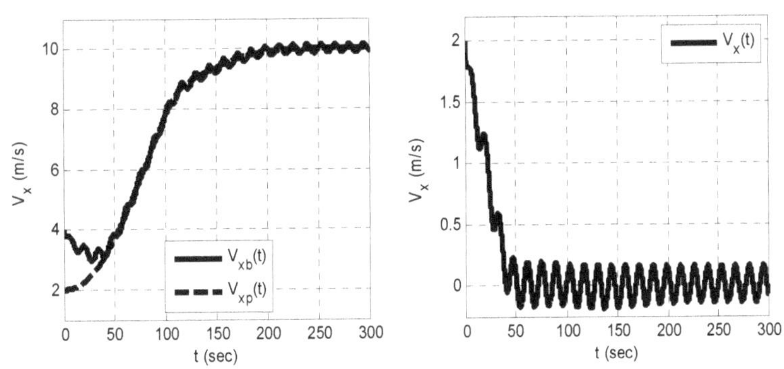

Figure 7. Forward speed $\overline{V}_x = V_{xb}(t), V_{x\rho}(t)$, and $V_x(t) = \overline{V}_x(t) - V_{x\rho}(t)$ under sea waves action.

Finally, let us analyze the stability of the closed-loop system (17), (35), (41) with synthesized tracking controllers. One can easily see that in order to provide aforementioned analysis, it is sufficient to consider the zero-equilibrium stability for the closed-loop system (27), (33) presented in deflections from the reference motion.

It is obvious that systems (27), (33) have zero equilibrium, and to investigate its stability features, let us introduces the following Lyapunov function candidate:

$$V = V(\mathbf{x}) = \lambda^2 V_x^2 + \boldsymbol{\xi}^T \mathbf{S} \boldsymbol{\xi} = \mathbf{x}^T \mathbf{S}_1 \mathbf{x},\ \mathbf{S}_1 = \begin{pmatrix} \lambda^2 & 0 \\ 0 & \mathbf{S} \end{pmatrix},\ \mathbf{x} := \begin{pmatrix} V_x \\ \boldsymbol{\xi} \end{pmatrix}. \tag{42}$$

Here, the symmetrical matrix $\mathbf{S} > 0$ is a solution of the algebraic Riccati equation, which is used in the range of LQR controller (37) synthesis:

$$\mathbf{S} = \begin{pmatrix} 0.0107 & -0.243 & -0.102 & -0.141 \\ -0.243 & 7.04 & 2.59 & 2.97 \\ -0.102 & 2.59 & 1.27 & 1.22 \\ -0.141 & 2.97 & 1.22 & 2.98 \end{pmatrix}.$$

Let us notice that the introduced function $V(\mathbf{x})$ satisfy the relationships

$$\alpha_1(\|\mathbf{x}\|) \leq V(\mathbf{x}) \leq \alpha_2(\|\mathbf{x}\|),\ \forall \mathbf{x} \in E^5, \tag{43}$$

with the following K-class functions:

$$\alpha_1(\|\mathbf{x}\|) = \sigma_{\min}(\mathbf{S}_1) \cdot \|\mathbf{x}\|^2,\ \ \alpha_2(\|\mathbf{x}\|) = \sigma_{\max}(\mathbf{S}_1) \cdot \|\mathbf{x}\|^2,$$

where $\sigma_{\min} = 0.00115$ and $\sigma_{\max} = 100$ are the minimum and maximum eigenvalues of the matrix $\mathbf{S}_1$, respectively.

Using the additional function $\alpha_3(\|\mathbf{x}\|) \equiv \|\mathbf{x}\| \in K$, one can check if the following inequality is valid

$$W(t,\mathbf{x}) = \left.\frac{dV}{dt}\right|_{(27),(33)} \leq -\alpha_3(\|\mathbf{x}\|),\ \forall t \geq 0,\ \forall \mathbf{x} \in B_x, \tag{44}$$

where B_x is a box, determined by the relationships:

$$B_x(t) = \left\{ \mathbf{x} \in E^5 : |x_i| \leq x_{im}(t),\ i = \overline{1,5} \right\},$$

$$x_{1m}(t) = 0.01P_\rho|V_{x\rho}(t)|,\ x_{2m}(t) = 0.01P_\rho|V_{y\rho}(t)|, x_{3m}(t) = 0.01P_\rho|\omega_\rho(t)|,\ x_{4m}(t) = 0.01P_\rho|\varphi_\rho(t)|,$$

$$x_{5m}(t) = 0.01P_\rho|\delta_\rho(t)|.$$

Here, the variable P_ρ determines the relative width of the box B_x compared to the current values of the reference signals. This variable was increased to such a value that the condition (44) was met. The obtained value $P_\rho = 25\%$ seems to be still admissible in the range of (43), (44) [4,5], determining the region of the local uniform asymptotic stability for the reference motion, which is realized by tracking controllers (35), (41).

4. Discussion

The main goal of this work was to propose constructive methods for marine tracking controllers' design taking into account the real conditions of a vessel's motions. We focused our main attention on a situation where the rudders' deflections and the forward speed are presented by initially given reference signals to be realized using tracking feedback controls.

This problem can be solved using different popular optimization approaches (Bellman's theory, MPC technique, sliding mode control, etc.). Nevertheless, we propose a new specific method for tracking controllers' design, which ensures the desirable reference motion of the vessel along the forward speed and heading angle.

This method is based on the optimal damping concept, which has certain advantages related to the practical requirements for the dynamic features of a closed-loop connection. The main advantage of the aforementioned approach is that the numerical solution of the optimization problems is essentially simplified. In contrast to well-known approaches [1–4], we applied OD tracking feedback as a control law with special features that allow it to be adjusted and implemented in real-time regime of motion.

The main result of this study is the development of the optimal damping concept to ensure its practical applicability and effectiveness that is illustrated by a controller design for a transport ship.

The investigations presented above could be further developed to consider the robust features of the tracking control laws, information about the measurement noise, and the presence of transport delays [24]. The results of the executed research could also be implemented to provide desirable reference motion of the dynamic positioning systems for marine vessels [26,27]. Certain attention may be given to the multipurpose control laws applications [22–25]. As for direct development of this study, it is possible to also use OD controller for the vertical rudders. The extension of the proposed approach to various robotic systems is also of considerable interest. The scope of the proposed approach may additionally include remotely operated vehicles [11] and offshore structures [12].

Funding: This research was funded by the Russian Foundation for Basic Research (RFBR) controlled by the Government of Russian Federation, research project number 20-07-00531.

Acknowledgments: The author expresses his sincere gratitude to A.P. Zhabko for the interest to the topic of this study and for the high-level professional suggestions.

Conflicts of Interest: The authors declare no conflict of interest.

References

1. Fossen, T.I. *Guidance and Control of Ocean Vehicles*; John Wiley & Sons Ltd.: New York, NY, USA, 1994.
2. Do, K.D.; Pan, J. *Control of Ships and Underwater Vehicles. Design for Underactuated and Nonlinear Marine Systems*; Springer: London, UK, 2009.
3. Jarjebowska, E. *Model-Based Tracking Control of Nonlinear Systems*; CRC Press, Taylor & Francis Group: Boca Raton, FL, USA, 2012.
4. Khalil, H.K. *Non-Linear Systems*; Prentice Hall: Englewood Cliffs, NJ, USA, 2002.
5. Slotine, J.; Li, W. *Applied Nonlinear Control*; Prentice Hall: Englewood Cliffs, NJ, USA, 1991.
6. Fossen, T.I. *Handbook of Marine Craft Hydrodynamics and Motion Control*; John Wiley & Sons, Ltd.: New York, NY, USA, 2011.
7. Reyes-Bayes, R.; Donaire, A.; van der Shaft, A.; Jayawardhana, B.; Perez, T. Tracking Control of Marine Craft in the port-Hamiltonian Framework: A Virtual Differential Passivity Approach. In Proceedings of the 17th European Control Conference (ECC 2019), Naples, Italy, 25—28 June 2019; pp. 1636–1641.
8. Liu, Y.; Bu, R.; Gao, X. Ship Trajectory Tracking Control Systems Design Based on Sliding Mode Control Algorithm. *Pol. Marit. Res.* **2018**, *25*, 26–34. [CrossRef]
9. Guerreiro, B.J.; Silvestre, C.; Cunha, R.; Pascoal, A. Trajectory tracking nonlinear model predictive control for autonomous surface craft. *IEEE Trans. Control. Syst. Techol.* **2014**, *22*, 2160–2175. [CrossRef]
10. Hammound, S. Ship Motion Control Using Multi-Controller Structure. *J. Marit. Res.* **2012**, *55*, 184–190.
11. Capocci, R.; Dooly, G.; Omerdi'c, E.; Coleman, J.; Newe, T.; Toal, D. Inspection-Class Remotely Operated Vehicles—A Review. *J. Mar. Sci. Eng.* **2017**, *5*, 13. [CrossRef]
12. Ramos, R.L. Linear Quadratic Optimal Control of a Spar-Type Floating Offshore Wind Turbine in the Presence of Turbulent Wind and Different Sea States. *J. Mar. Sci. Eng.* **2018**, *6*, 151. [CrossRef]
13. Roy, S.; Shome, S.N.; Nandy, S.; Ray, R.; Kumar, V. Trajectory Following Control of AUV: A Robust Approach. *J. Inst. Eng. India Ser. C* **2013**, *94*, 253–265. [CrossRef]
14. Ye, J.; Roy, S.; Codjevac, M.; Baldi, S. A Switching Control Perspective on the Offshore Construction Scenario of Heavy-Lift Vessels. *IEEE Trans. Control Syst. Technol.* **2020**, *29*, 1–8. [CrossRef]
15. He, W.; He, X.; Zoo, M.; Li, H. PDE Model-Based Boundary Control Design for a Flexible Robotic Manipulator With Input Backlash. *IEEE Trans. Control Syst. Technol.* **2019**, *27*, 790–797. [CrossRef]
16. He, W.; Gao, H.; Zhou, C.; Yang, C. Reinforcement Learning Control of a Flexible Two-Link Manipulator: An Experimental Investigation. *IEEE Trans. Syst. Man Cybern. Syst.* **2020**. [CrossRef]
17. Lewis, F.L.; Vrabie, D.L.; Syrmos, V.L. *Optimal Control*; John Wiley & Sons, Ltd.: New York, NY, USA, 2012.
18. Geering, H.P. *Optimal Control with Engineering Applications*; Springer-Verlag: Berlin/Heidelberg, Germany, 2007.
19. Zubov, V.I. *Oscillations in Nonlinear and Controlled Systems*; Sudpromgiz: Leningrad, USSR, 1962. (In Russian)
20. Zubov, V.I. *Theory of Optimal Control of Ships and Other Moving Objects*; Sudpromgiz: Leningrad, USSR, 1966. (In Russian)
21. Zubov, V.I. *Theorie de la Commande*; Mir: Moscow, USSR, 1978.
22. Veremey, E.I. Synthesis of multiobjective control laws for ship motion. *Gyroscopy Navig.* **2010**, *1*, 119–125. [CrossRef]
23. Veremey, E.I. Dynamical Correction of Control Laws for Marine Ships' Accurate Steering. *J. Mar. Sci. Appl.* **2014**, *13*, 127–133. [CrossRef]
24. Veremey, E.I. Optimization of filtering correctors for autopilot control laws with special structures. *Optim. Control Appl. Methods* **2016**, *37*, 323–339. [CrossRef]
25. Veremey, E.I. Special Spectral Approach to Solutions of SISO LTI H-Optimization Problems. *Int. J. Autom. Comput.* **2019**, *16*, 112–128. [CrossRef]
26. Veremey, E.I. Separate Filtering Correction of Observer-Based Marine Positioning Control Laws. *Int. J. Control* **2017**, *90*, 1561–1575. [CrossRef]
27. Sotnikova, M.V.; Veremey, E.I. Dynamic Positioning Based on Nonlinear MPC. *IFAC Proc. Vol. (IFAC Pap.)* **2013**, *9*, 31–36. [CrossRef]
28. Veremey, E.I.; Pogozhev, S.V.; Sotnikova, M.V. Multipurpose Control Laws Synthesis for Actuators Time Delay. *J. Mar. Sci. Eng.* **2020**, *8*, 477. [CrossRef]
29. Chen, M.; Ge, S.S.; How, B.V.; Choo, Y.S. Robust adaptive position mooring control for marine vessels. *IEEE Trans. Control Syst. Technol.* **2013**, *21*, 395–409. [CrossRef]

Journal of
Marine Science and Engineering

MDPI

Article

Research on Early Warning of Ship Danger Based on Composition Fuzzy Inference

Zhiying Guan, Yan Wang, Zheng Zhou and Hongbo Wang *

State Key Laboratory on Integrated Optoelectronics, College of Electronic Science and Engineering, Jilin University, Changchun 130000, China; guanzy18@mails.jlu.edu.cn (Z.G.); yanwang20@mails.jlu.edu.cn (Y.W.); zhouzheng20@mails.jlu.edu.cn (Z.Z.)

* Correspondence: wang_hongbo@jlu.edu.cn; Tel.: +86-0431-85148242

Received: 10 November 2020; Accepted: 4 December 2020; Published: 8 December 2020

Abstract: Ship collision avoidance measures are important for reducing marine accidents caused by human factors and various natural environmental factors and can also prevent property loss and casualties. In recent years, various methods have been used to study collision avoidance, including ship domain models. This paper proposes a ship domain model based on fuzzy logic aimed at providing early warning of ship collision risk and a reasonable reference that can be used in combination with the International Regulation for Preventing Collisions at Sea (COLREGs). The composition fuzzy inference combining more than one fuzzy inference process is first used to introduce as many factors as possible related to ship collision risk for calculating the ship domain. In this way, the calculation of the ship domain size is more accurate, and a more accurate reference can be provided to sailors, which could save both time and labor by reducing errors. A fuzzy inference system based on if-then fuzzy rules was established in MATLAB and simulation experiments were conducted. The simulation results suggest that the proposed method is feasible and can help sailors make subjective decisions to effectively avoid the occurrence of collision accidents.

Keywords: collision avoidance; ship domain; fuzzy inference; collision risk; early warning system

1. Introduction

In recent years, with advances in science and technology, maritime traffic has increased and brought many conveniences to human life. However, the increasing complexity of maritime traffic has gradually led to more safety problems. Frequent ship collisions cause serious economic losses and threaten the safety of passengers and crew members. Furthermore, worsening environmental pollution caused by ship collisions cannot be neglected. Despite a large body of research and methods devoted to reducing collisions at sea, ship collision accidents are inevitable. Thus, ship collision avoidance has become a research hotspot.

To minimize ship collisions, thereby increasing the safety of ships sailing at sea and reducing the associated losses, researchers in various countries have explored a range of methods for ship collision avoidance. According to the European Maritime Safety Agency's review [1] of maritime casualties in 2017, EU member states reported 3145 maritime accidents in 2016. More than half of these accidents were caused by ship collisions, and more than 60% of the collisions were caused by human error. Thus, reducing human errors during navigation at sea has become an important focus of research, with an emphasis on improving the navigator decision-making processes.

Algorithms such as the genetic algorithm, neural network, and distributed algorithm have been applied to ship collision avoidance problems. In addition, automatic radar plotting aid (ARPA) and automatic identification system (AIS) have been used to help make collision avoidance decisions. ARPA offers accurate information about the bearing and distance of nearby obstacles, while AIS provides important information about the ship's course, speed, and position. These data provide a powerful basis for effective ship collision avoidance. By far, many researchers have studied the ship domain in different waters with the help of AIS data. For example, Xiang et al. [2] proposed a calculation method for ships in restricted waters based on AIS data. Hansen M.G. et al. [3] used AIS data to study the ship domain in the bridge area and crossing waters and Zhang et al. [4] develop an effective big AIS data-driven approach to determine the probabilistic ship domain. Ding et al. [5] used AIS data to make statistics on the ship domain of different types of ships and ships of different sizes in open waters. Although research using AIS is now common, these systems can only be used as assistant tools to identify obstacles and offer additional navigation information. While these tools provide accurate and reliable information for navigators as the basis for decision-making, sailors cannot use this information to make navigation decisions.

Some algorithms have been used to aid ship navigation decision-making systems. For example, Li et al. [6] used an improved multiobjective algorithm to plan the collision avoidance trajectory of ships and to suggest safe navigation routes. In addition, Li et al. [7] studied ship collision avoidance using distributed algorithms for preventing multi-ship collisions at sea. Tsou and Hsueh [8] combined the ant colony algorithm with onboard AIS and geographic information system (GIS) systems for safe and economical collision avoidance route planning. The algorithms mentioned above, and others, can be used to plan the general ship collision avoidance route for multiple ships in the same water area, but the management of temporary accidents has not been well studied.

Previously, there have been researchers who used fuzzy logic to study navigation issues. Lee et al. [9] proposed a fuzzy logic ship autonomous navigation algorithm based on COLREGs criterion. The VFF (Virtual Force Field) method is improved and applied to the autonomous navigation of ships. However, the fuzzy inference system is mainly based on COLREGs, which lacks sufficient consideration of the interference caused by various factors in navigation. Grinyak et al. [10] proposed a fuzzy decision-making system about the motion's danger level that combines Mamdani and Sugeno fuzzy inference systems, but it mainly considered a model of the relative motion of two vessels and neglected the environmental factors. Namgung [11] proposed a fuzzy inference system that expressed appropriate collision risk index corresponding to level. It can measure the fuzzy risk index, but it cannot play a direct role in the sailor's decision warning. In this paper, composition fuzzy inference is carried out by comprehensively considering the ship's own factors and environmental factors when the ship is sailing at sea, giving the ship early collision warning and reminding the sailor to take measures to avoid collision.

When malfunction occurs or the ship fails to sail along an expected path or at an expected speed, the sailor's timely response is crucial. In the past, sailors' decisions were based on experience but with further research on the behavior of the navigating ships and intelligence science and technology, collision risk warnings can be achieved by other means. For instance, the ship domain model can be used to evaluate collision risk. Ship domains represent a defined area around the ship that other ships should not enter to ensure that the ship navigates safely through waters. Thus, the ship domain can be used as a standard safety measure and a reference parameter for evaluating collision risk, and it is also the basis for taking actions in ship collision avoidance.

The concept of ship domains was first proposed by Japanese scholar Fujii [12] as an elliptical domain proportional to the length of the ship. Goodwin [13], Davis [14], Coldwell [15], and others were inspired to establish other ship domain models' statistical observations. Vander Tak et al. [16] used the ship domain model to calculate the frequency of ship encounters for assessing maritime traffic hazards. Later, the ship domain concept was further extended. Pietrzykowski and Uriasz [17–19] introduced the notion of the fuzzy ship domain and defined ship domains in both open waters and restricted waters. Qu et al. [20] proposed three ship collision risk indicators based on the ship domain to quantitatively evaluate ship collision risk in a strait. Wang et al. [21] proposed a fuzzy quaternion ship domain model (FQSD) and used analytic expressions to describe models. Zhou et al. [22] introduced the dynamic fuzzy ship domain and established a ship domain model to determine collision risk between ships.

In this paper, the ship domain is used as the criterion for assessing collision risk. When other ships enter the ship domain, the situation becomes urgent and the ship must take action to avoid an imminent collision. Therefore, it is necessary to study the range and size of the ship domain. This paper proposes a fuzzy logic approach. The size of the ship domain is regarded as a fuzzy value that varies with the external environmental factors. A fuzzy control method is adopted to obtain a reasonable value of the ship domain by considering the dangerous inviolable area of each ship. Through fuzzification, fuzzy reasoning, and defuzzification, a reference for collision avoidance between ships is provided to effectively assist navigational decision-making.

In view of the fact that there are many factors that affect the ship domain, a single fuzzy inference may not be able to consider as many influencing factors as possible, which limits the number of factors. So, the factors considered in the study of ship collision warning were always limited in the past. Moreover, environmental factors such as navigation density and fairway visibility are always ignored. However, removing too many factors from the model can lead to excessive errors. The proposed composition fuzzy inference method combines the results of multiple fuzzy inferences and then performs fuzzy inference, retaining as many relevant factors as possible, which solves the problem of oversize errors caused by too many factors that need to be deleted in the past, thereby providing a more accurate estimate of the ship domain size, which can be used for more accurate and timely warning of collision danger to reduce the occurrence of collisions.

In addition, the previous research can only obtain the relationships between the ship domain and certain single factors. Here, the fuzzy logic method was used to not only obtain the changes in the ship domain with certain factors but also to simulate the influence of multiple factors on the size of the ship domain, which enables a more accurate description of the ship domain size. The proposed ship domain model could save both time and labor and enable the ship domain concept to be better applied in navigation systems.

The rest of the paper is organized as follows. Section 2 describes the ship domain and its influencing factors, various factors are selected to determine the shape of the ship domain, and less significant factors are removed. In Section 3, a fuzzy inference model of the ship domain is established considering five influencing factors including both own ship factors and environmental factors. In Section 4, the simulation platform and inference system are introduced and the simulation experiments and results compared with the traditional methods are presented. In Section 5, we summarize the advantages of the proposed method and the importance of early warning systems for preventing ship collisions.

2. Ship Domain Model and Influencing Factors

2.1. Ship Domain

Fujii [12] defined ship domain as "The area around the previous ship most of the following ship's sailors avoid entering." The dimensions of the ship domain are related to factors such as ship speed, density, and tidal current. The first established ship domain model was oval shaped, as shown in Figure 1. Fujii assigned a specific numerical value to the ship domain size based on observations of maritime traffic, making it no longer an abstract concept, and demonstrated that the range of ship domain varies with the navigation conditions and waters. Later, other scholars introduced ship domains, which were mainly elliptical [2–5]. Therefore, in the present, an elliptical ship domain was selected.

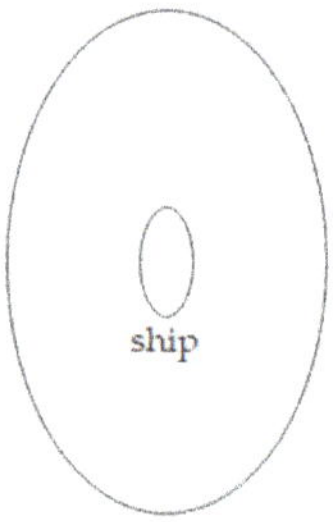

Figure 1. Fujii's ship domain model.

2.2. Factors Affecting the Ship Domain

The ship domain depends on the sailor's maneuvering, however, the sailor's response is based on objective facts and also influenced by various factors. Therefore, the ship domain is affected by many subjective and objective uncertainties, including human, ship, environment, and management factors, and changes dynamically. The ship domain may also vary with its factors for different research purposes and different types of waters under different environmental conditions. The main factors are as follows:

(1) Weather and hydrological effects. Generally, reduced visibility increases the size of the ship domain.
(2) Ship size. The larger the ship, the larger the ship domain.
(3) Vessel speed. The larger the ship speed, the larger the ship domain.
(4) Surrounding waters around which the ship is sailing affect the size of the ship domain. The ship domain is large in open waters and small in restricted water.
(5) Maritime navigation density. Higher navigation density in the vessel's sailing area results in a smaller ship domain.
(6) Route conditions. For example, when the vessel is navigating a curved channel or jet zone, the domain is large.
(7) Maneuverability of the vessel. Improved maneuverability decreases the ship domain.
(8) Skill of the sailor.
(9) Encounter angle between two encountering vessels.
(10) Types of vessels.

The influence of various uncertain factors on the ship domain is complex, which makes it difficult to determine the relationships between ship domain and each factor. The influencing factors in the ship domain must, therefore, be analyzed and factors that have an obvious influence on the size of the ship domain should be selected, which can significantly reduce the complexity of studies of the ship domain.

At present, research on the ship domain is mostly aimed at open waters or restricted waters, and ship domains in diverse waters are suitable for different situations. In this study, we consider open waters. Management factors are not considered since they are not highly relevant in opens waters. Zhou's work [22] shows that the type of ship has little effect on the size of the ship domain, and its impact is mainly reflected in the speed of the vessel. Thus, the impact of ship type will not be considered in this study to simplify the ship domain model.

According to previous analyses with multiple sample data, among the many factors affecting the size of the ship domain, there is a clear trend in size, speed, and encounter angle of the ship. The size of the ship domain is also affected by navigation density and fairway visibility, whereas the sailor class and ship type have little effect on the size of the ship domain and no obvious rules have been defined. This paper considers three own ship factors, including ship length, ship speed, and encounter angle between two ships, and two environmental factors, including maritime navigation density and fairway visibility.

3. Composition Fuzzy Inference Based on Ship Domain

Herein, the influence of various factors on the size of the ship domain are determined. The shape of the ship domain is not discussed. The ship domain was assumed to be an ellipse. Since the position of the ship inside the domain is different in every study, only the size of the ship domain is assessed. The lengths of the major and the minor axes of the ellipse can determine its area. The situation is similar when these lengths are used for fuzzy control. However, the major axis length could better predict collision risk and was therefore selected as the output variable, indicating the impact of each factor on the size of the ship domain.

In this paper, five factors influencing the ship domain are considered: ship length, ship speed, encounter angle between the two ships, navigation density, and fairway visibility. The excessive number of factors make single fuzzy reasoning impossible because the formulation of excessive fuzzy rules is extremely complicated and may produce large errors. Therefore, the composition fuzzy reasoning method is adopted. First, the influence of the ship's own factors on the size of the ship domain is simulated through fuzzy inference to form ship domain model D1. Then, the environmental factors are simulated to form ship domain model D2. Finally, models D1 and D2 are combined. After fuzzy inference, ship domain model D can be defined as the ship domain model simultaneously affected by the ship's own factors and environmental factors, as shown in Figure 2.

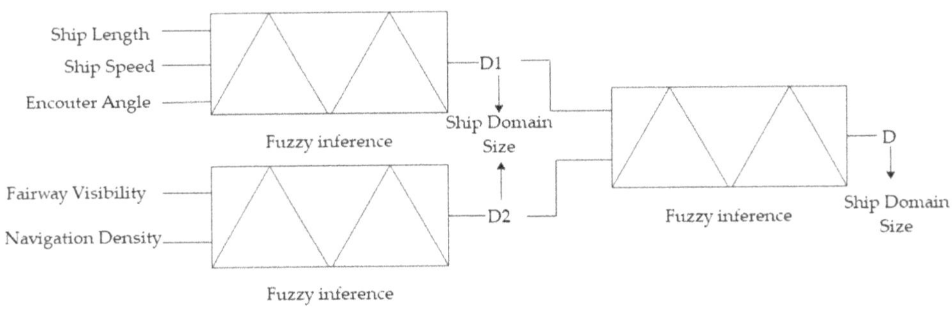

Figure 2. Composition fuzzy inference model.

3.1. *Fuzzy Ship Domain Affected by Ship's Own Factors*

3.1.1. Fuzzy Control

Fuzzy control is the process that performs fuzzy inference on input variables to transform the fuzzy output values into crisp output values using the following steps:

(1) Select the input and output variables that reflect the working mechanism of the system and determine the range of input and output variables and the fuzzy set of linguistic variables;
(2) Create the membership function (FMF) for each set of inputs and outputs;
(3) Formulate the if-then fuzzy rules to construct the relationship between the output fuzzy set and the input fuzzy sets, and select the fuzzy reasoning model;
(4) Perform fuzzy reasoning on the fuzzy input variables combined with the fuzzy rules and defuzzification of the output. Fuzzification converts the precise input value into a fuzzy value and defuzzification converts the fuzzy value after fuzzy inference into a crisp value using the following formulas, respectively:

$$x = fz(x_0) \tag{1}$$

$$z_0 = df(z), \tag{2}$$

where x_0 is the crisp input value, x is the fuzzy set, fz represents the fuzzification operator; z is the fuzzy value of the control output, z_0 is the clarity value of the control output, and df represents the defuzzification operator.

To consider the influence of different factors on the ship domain model, three critical factors are selected as the input linguistic variables, i.e., the ship's own size, the ship speed, and the encounter angle between the two ships. The major axis length of the elliptical ship domain is the output variable. Fuzzy control is performed to obtain a crisp output quantity of the size of the ship domain. A block diagram of the fuzzy control process of the ship domain model is shown in Figure 3.

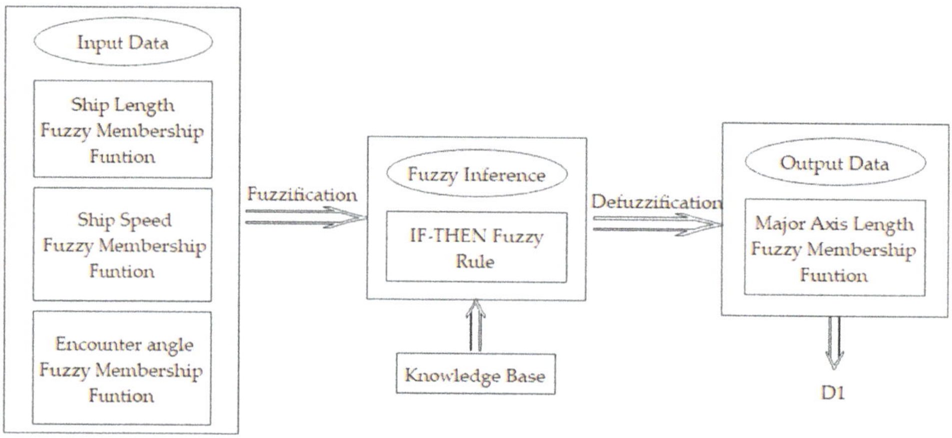

Figure 3. Block diagram of fuzzy control process for ship domain model influenced by ship's own factors.

3.1.2. Fuzzy Linguistic Variables and Fuzzy Sets

Fuzzy linguistic variables are fuzzy sets expressed in fuzzy language, simply referred to as linguistic variables, and can be divided into input linguistic variables and output linguistic variables. Each fuzzy linguistic variable has multiple fuzzy linguistic terms whose names usually have certain meanings, such as very small (VS), small (S), medium small (MS), medium (M), medium big (MB), big (B), very big (VB), etc.

The ship's own size, ship speed, and encounter angle between the two ships are selected as the input linguistic variables. The size of the ship domain is the quantity of fuzzy control selected as the output linguistic variable.

Each linguistic variable is composed of a set of fuzzy linguistic values, which constitutes a fuzzy set. For each linguistic variable, each fuzzy set of values has the same range.

The size of the ship itself can be represented by the length and breadth of the vessel. In this paper, "ship length" (L) is regarded as an input linguistic variable, which is divided into three fuzzy linguistic terms: small (S), medium (M), and big (B). That is to say, the values of the fuzzy set of the input linguistic variable "ship length" are {S, M, B}. By referring to the relevant data of ship size, the range of ship length is [30,400] and the unit of ship length is in meters (m). Similarly, after investigating typical ship speeds, the range of input linguistic variable "ship speed" (V) is [0,30] in kn. The values of the fuzzy set "ship speed" are also {S, M, B}, again representing small (S), medium (M), and big (B).

For the "encounter angle between the two ships" (E), the influence of the encounter angle on the size of the ship domain is more complex, therefore, more fuzzy linguistic terms must be selected, but it is also more precise. The input linguistic variable "encounter angle" is divided into seven fuzzy linguistic terms: very small (VS), small (S), medium small (MS), medium (M), medium big (MB), big (B), and very big (VB). That is, the values of the fuzzy set "encounter angle" are {VS, S, MS, M, MB,

B, VB}. The encounter angle is symmetrical to the ship, the bow direction is 0°, and the stern is 180°, which is determined as [0,180].

The output linguistic variable is the length of the long axis of the ship domain, represented as D1 here. The range of the output variable can be roughly determined by reviewing existing data on ship domain sizes. The range of output linguistic variable "ship domain size affected by ship's own factors" D1 is [150,3500] and the values of the fuzzy set are {S, MS, M, MB, B}, corresponding to five fuzzy linguistic terms: small (S), medium small (MS), medium (M), medium big (MB), and big (B).

3.1.3. Fuzzy Membership Functions

A linguistic variable has multiple linguistic terms and each linguistic term corresponds to a membership function. In the MATLAB Fuzzy Logic Toolbox, there are Gaussian, triangular, trapezoidal, bell-shaped, sigmoidal, л-type, and Z-type membership functions. Here, the triangular and trapezoidal membership functions are selected to describe the linguistic variables, with a, b, c, and d as the parameters, and the formulas are as follows, respectively:

$$f(x,a,b,c) = \begin{cases} 0, & x \le a \\ \frac{x-a}{b-a}, & a \le x \le b \\ \frac{c-x}{c-b}, & b \le x \le c \\ 0, & x \ge c \end{cases} \tag{3}$$

$$g(x,a,b,c,d) = \begin{cases} 0, & x \le a \\ \frac{x-a}{b-a}, & a \le x \le b \\ 1, & b \le x \le c \\ \frac{d-x}{d-c}, & c \le x \le d \\ 0, & x \ge d \end{cases}. \tag{4}$$

The parameters in the formulas are affected by the range of each linguistic value of the linguistic variable.

Figure 4 shows the membership functions of three input linguistic variables and one output linguistic variable. The input linguistic variable "ship speed" (V) selects only the triangular membership function, while the input linguistic variables "ship length" (L) "encounter angle between the two ships" (E), and the output linguistic variable "ship domain size" (D1) not only selects the triangular membership function but also uses the trapezoidal membership function. The Fuzzy Logic Toolbox in MATLAB was used to draw the membership functions, which not only reflects the membership function of the input and output linguistic variables but also clearly shows the linguistic term and range of input and output linguistic variables, as well as the relationship among each linguistic term of each linguistic variable.

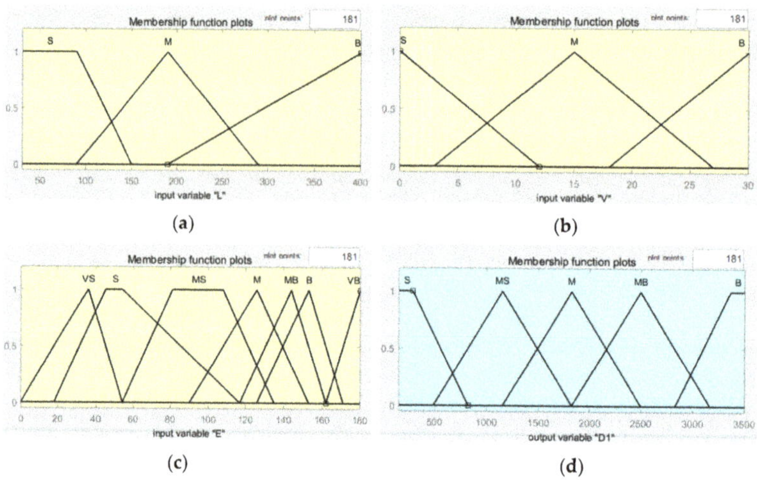

Figure 4. Fuzzy membership functions: (**a**) ship length (L), (**b**) ship speed (V), (**c**) encounter angle (E), and (**d**) ship domain size affected by ship's own factors (D1).

3.1.4. Fuzzy Inference

The fuzzy inference system applies fuzzy logic to map a given input to the output. In fuzzy inference systems, fuzzy rules can describe human experience and knowledge in the form of fuzzy language. A set of fuzzy logic inference systems is composed of many fuzzy inference rules. The number of fuzzy rules is the Cartesian product of the fuzzy linguistic variables and membership sets.

In the present study, there are three input linguistic variables. The input linguistic variable "ship length" has three membership sets, S, M, and B, "ship speed" also has three membership sets, S, M, and B, and "encounter angle" has seven membership sets, VS, S, MS, M, MB, B, and VB. Therefore, the number of fuzzy rules is 63. The fuzzy rules are presented in Table 1. The Mamdani model is the most commonly used fuzzy inference method and is adopted in the present study.

Table 1. Fuzzy rules for ship domain size (D1).

Ship Domain Size (D1)		**Encounter Angle (E)**						
Ship Length (L)	**Ship Speed (V)**	**VS**	**S**	**MS**	**M**	**MB**	**B**	**VB**
S	S	MB	M	M	MS	MS	MS	S
	M	MB	MB	M	M	MS	MS	S
	B	B	MB	MB	M	M	MS	MS
M	S	B	MB	M	M	MS	MS	S
	M	B	B	MB	M	M	MS	MS
	B	B	B	MB	MB	M	M	MS
B	S	B	MB	MB	M	M	MS	MS
	M	B	B	MB	MB	M	M	MS
	B	B	B	B	MB	MB	M	MS

After the input and output linguistic variables are selected, their ranges and fuzzy sets are determined. The membership functions of the input and output linguistic variables are also selected as well as the fuzzy reasoning model. The MATLAB Fuzzy Logic Toolbox was used to establish the fuzzy inference system, taking the ship length (L), ship speed (V), and encounter angle (E) as the input variables and the ship domain size (D1) as the output variable.

3.2. *Influence of Environmental Factors on Fuzzy Ship Domain*

3.2.1. Fuzzy Control Process

The influence of environmental factors on the fuzzy ship domain affected are also considered. Two factors, navigation density and fairway visibility, are selected as input linguistic variables and the long axis length of the elliptical ship domain is taken as an output linguistic variable. Fuzzy reasoning can be performed to obtain the ship domain size. A block diagram of the fuzzy control process for the ship domain model is shown in Figure 5.

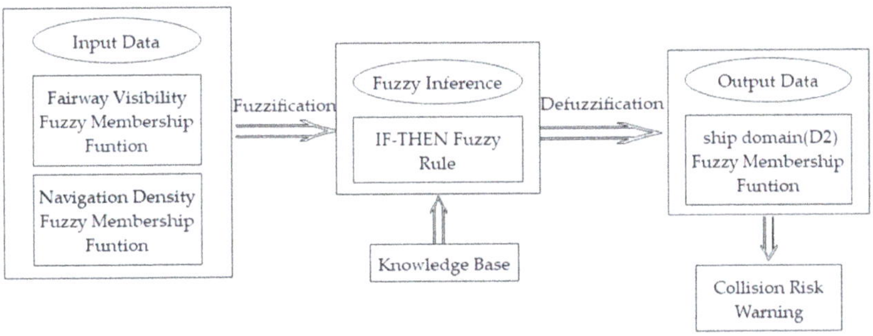

Figure 5. Block diagram of fuzzy control process for ship domain model affected by environmental factors.

3.2.2. Fuzzy Linguistic Variables and Fuzzy Sets

In this paper, "navigation density" N is divided into five fuzzy linguistic terms: small (S), medium small (MS), medium (M), medium big (MB), and big (B). Therefore, the values of the fuzzy set of the input linguistic variable "navigation density" N are {S, MS, M, MB, B}. After consulting relevant data on navigation density, the range of navigation density is determined as [0,600], referring to the number of ships passing per unit time. Generally, 24 h is taken as the unit time.

Similarly, after investigating the visibility in a certain area, the range of the input linguistic variable "fairway visibility" K is [0,25] in km. The values of the fuzzy set "fairway visibility" are {S, MS, M, MB, B}, which represent the five fuzzy linguistic terms of fairway visibility, respectively, small (S), medium small (MS), medium (M), medium big (MB), and big (B). For the output linguistic variable, "ship domain size affected by environmental factors" D2 is selected to distinguish from the ship domain size D1 influenced by the ship's own factors. The range of the ship domain size (D2) is consistent with that of the ship domain size affected by ship's own factors (D1), which is [150,3500], and the values of the fuzzy set are {S, MS, M, MB, B}.

3.2.3. Fuzzy Membership Functions

The triangular membership function and trapezoidal membership function are again selected to describe the linguistic variables, as shown in Figure 6.

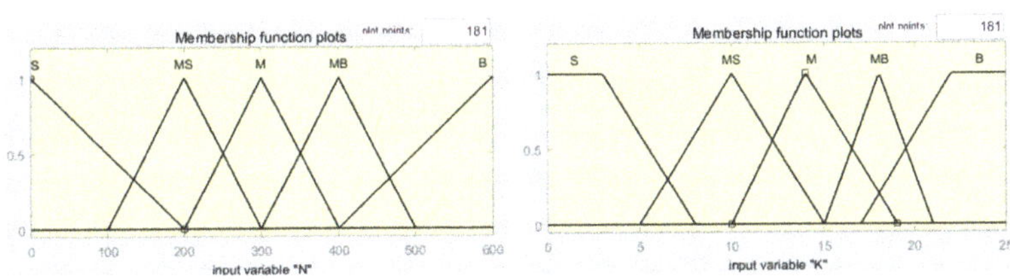

(a) (b)

Figure 6. *Cont.*

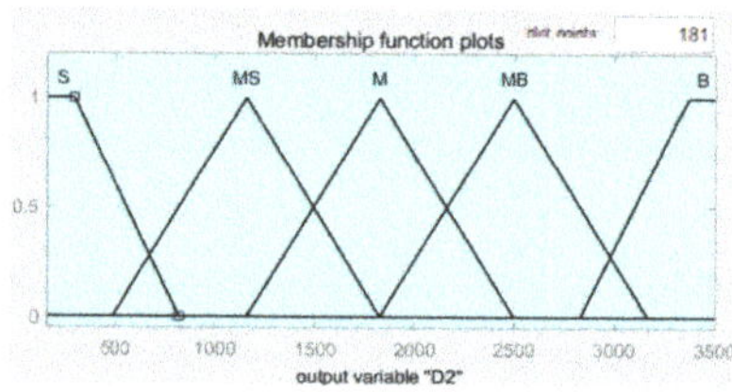

(c)

Figure 6. Fuzzy membership functions: (**a**) navigable density (N), (**b**) fairway visibility (K), and (**c**) ship domain size by environmental factors (D2).

3.2.4. Fuzzy Inference

The fuzzy rules of navigation density, fairway visibility and ship domain size are shown in Table 2, and the Mamdani fuzzy reasoning method is adopted.

Table 2. Fuzzy rules for ship domain size (D2).

Ship Domain Size (D2)	Navigation Density (N)				
Fairway visibility (K)	S	MS	M	MB	B
S	B	B	B	MB	MB
MS	B	MB	MB	M	MS
M	MB	M	M	M	MS
MB	MB	M	MS	MS	S
B	MS	MS	S	S	S

3.3. *Ship Domain of Composition Fuzzy Inference*

3.3.1. Fuzzy Control Process

The fuzzy ship domain is affected by both the ship's own factors and environmental factors. Using the composition inference method, ship domain size affected by the ship's own factors (D1) and ship domain size affected by environmental factors (D2) can be obtained and are taken as the input linguistic variables. The long axis size of the oval ship domain is taken as the output linguistic variable, expressed as D. In other words, under composition fuzzy inference, the size of ship domain is affected by both ship's own factors and environmental factors. A block diagram of the fuzzy control process for the ship domain model is presented in Figure 7.

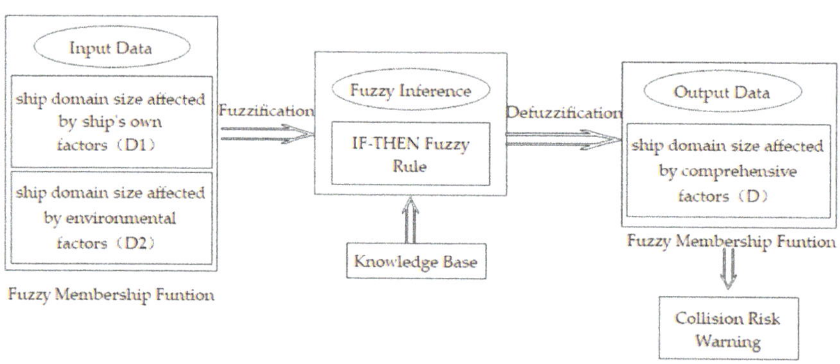

Figure 7. Block diagram of fuzzy control process for ship domain model.

3.3.2. Fuzzy Linguistic Variables and Fuzzy Sets

Both the input linguistic variable and the output linguistic variable are the ship domain size; "ship domain size affected by the ship's own factors" D1 and "ship domain size affected by environmental factors" D2 are the input linguistic variables; and "the ship domain size affected by comprehensive factors" D, which is influenced simultaneously by both the ship's own factors and environmental factors, is the output linguistic variable.

Since the input and output linguistic variables both represent values of the ship domain long axis size, their range and fuzzy linguistic terms are the same and can be divided into five fuzzy linguistic terms: small (S), medium small (MS), medium (M), medium big (MB), and big (B). The values of the fuzzy set are {S, MS, M, MB, B}. The range is [150,3500].

3.3.3. Fuzzy Membership Functions

The membership function is shown in Figure 8.

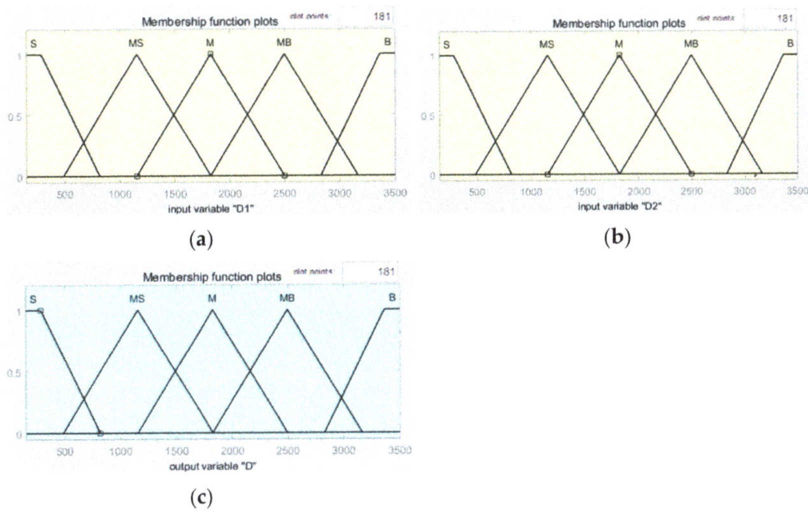

Figure 8. Fuzzy membership functions: (**a**) ship domain size affected by ship's own factors (D1), (**b**) ship domain size by environmental factors (D2), and (**c**) ship domain size affected by comprehensive factors (D).

3.3.4. Fuzzy Inference

The fuzzy rules are presented in Table 3. The Mamdani fuzzy reasoning method is adopted.

Table 3. Fuzzy rules (D).

Ship Domain Size Affected by Comprehensive Factors (D)	Ship Domain Size Affected by Environmental Factors (D2)				
Ship domain size affected by the ship's own factors (D1)	S	MS	M	MB	B
S	S	S	S	MS	M
MS	S	MS	MS	MS	M
M	MS	M	M	M	MB
MB	M	MB	MB	MB	B
B	M	MB	B	B	B

4. Results

The MATLAB Fuzzy Logic Toolbox was used to construct and then simulate the fuzzy inference system, then the relationships between the size of the ship domain and influencing factors, including ship length (L), ship speed (V), encounter angle (E), navigation density (N), and fairway visibility (K) were obtained, as shown in Figure 9.

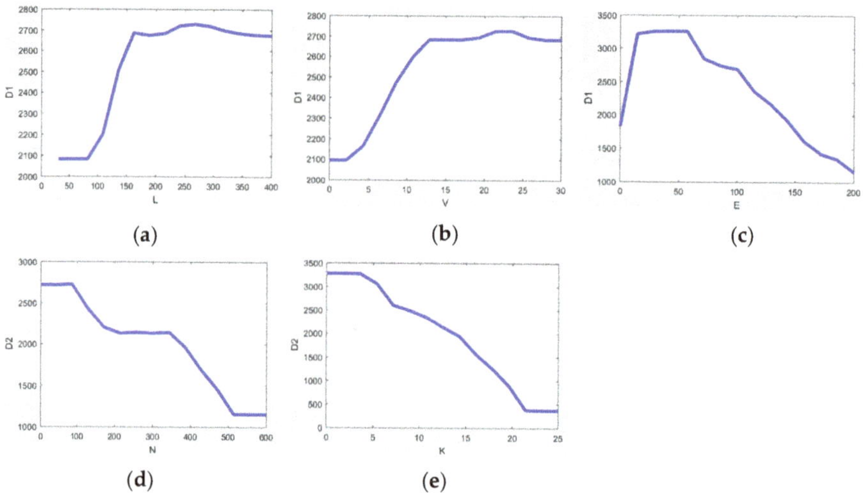

Figure 9. Variation of ship domain size with influencing factors: (**a**) ship length (L), (**b**) ship speed (V), (**c**) encounter angle (E), (**d**) navigation density (N), and (**e**) fairway visibility (K).

The ship domain shows a trend of gradually increasing with the increase in size and speed of the ship, but gradually decreases with increasing encounter angle. Simultaneously, the size of the ship domain gradually decreases with increasing navigation density and fairway visibility, i.e., the smaller the navigation density, the lower the visibility and the larger the ship domain size. However, the ship domain does not have a linear relationship with these single factors. Thus, it can be seen, the influence result of each factor on ship domain obtained by fuzzy inference is similar to other studies [22].

The comprehensive influence of various two-factor combinations on the size of the ship domain were also simulated, as shown in Figure 10. The surface diagrams illustrate the trends of each set of influencing factors on the size of the ship domain as well as the specific size of the ship domain when concrete values of the two factors are input into the MATLAB simulation platform.

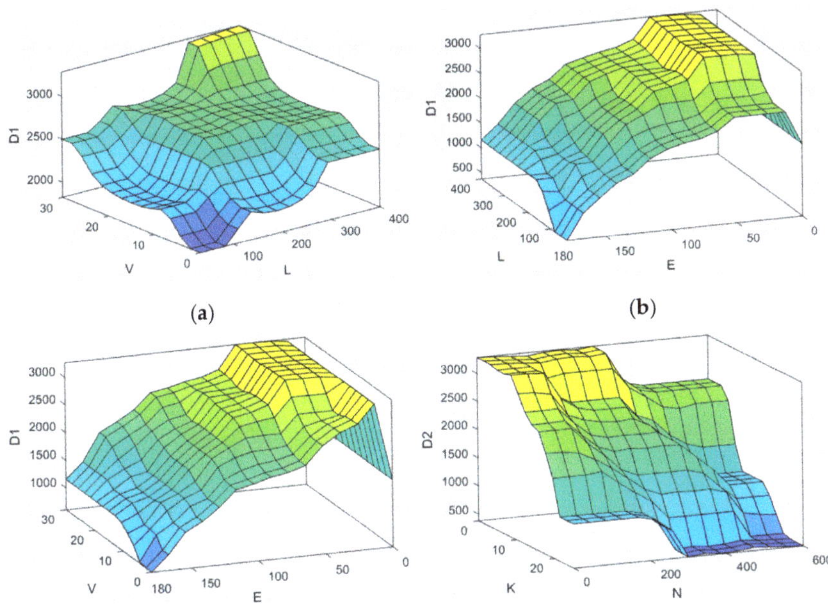

(a) (b) (c) (d)

Figure 10. Comprehensive influence of various two-factor combinations on the size of the ship domain: (**a**) ship length (L) and ship speed (V), (**b**) ship length (L) and encounter angle (E), (**c**) ship speed (V) and encounter angle (E), and (**d**) navigation density (N) and fairway visibility (K).

The MATLAB Fuzzy Logic Toolbox was used for simulations and the specific size of the ship domain was obtained by determining the input. The results of the fuzzy simulations are shown in Figure 11. When specific values are assigned as inputs, i.e., the values of ship length (L), ship speed (V), encounter angle (E), navigation density (N), and fairway visibility (K) are determined, the size of the ship domain can be obtained. This article makes the simulation when two ships are in the head-on situation as an example, when the "ship length" L = 215 m, the "ship speed" V = 15 kn, the "encounter angle" E = 5, "navigation density" N = 420, and "fairway visibility" K = 10 km, and "the major axis length of the elliptical ship domain" D = 2990 m. If the ship does not encounter any other ships at this time, appropriate actions should be taken according to the length of the semimajor axis and the International Regulation for Preventing Collisions at Sea (COLREGs).

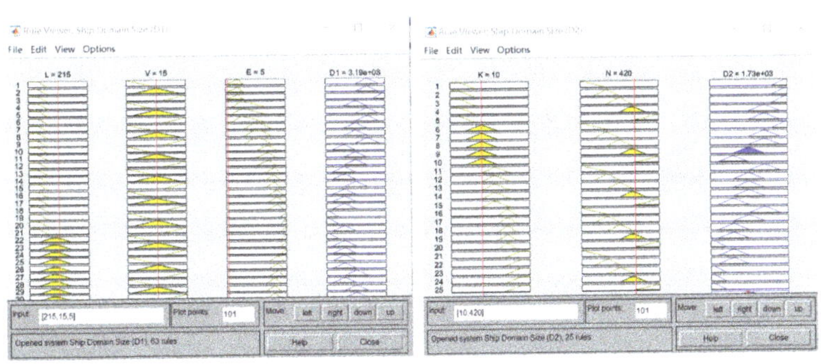

(a) (b)

Figure 11. *Cont.*

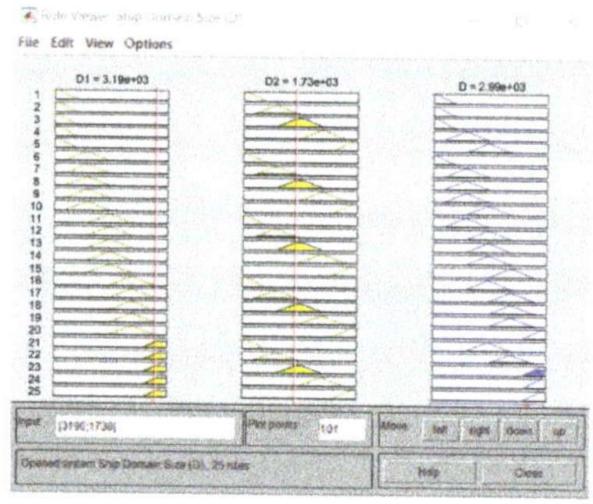

(c)

Figure 11. Fuzzy simulation results: (**a**) ship domain size affected by ship's own factors (D1), (**b**) ship domain size by environmental factors (D2), and (**c**) ship domain size affected by comprehensive factors (D).

Figure 12 shows the ship domain of the own vessel and the encounter situation between the own vessel and a target vessel sailing in one direction. Taking the major axis of the ellipse as the diameter and the center of the ship as the center of the ship domain, a circle tangent to the ellipse was drawn. From Figure 12, it can be easily observed that when the "encounter angle" E = 5, the two ships are in the head-on situation, the own vessel receives a collision risk warning when the target vessel is 1495 m away and collision avoidance actions according to COLREGs should be taken to avoid an urgent situation.

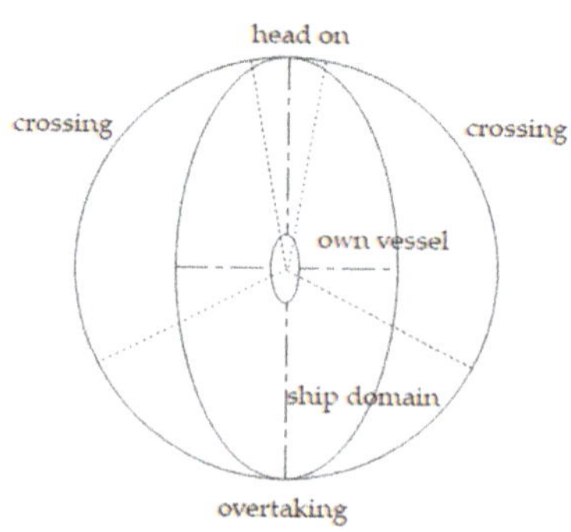

Figure 12. Ship encounter situation.

In other encounter situations, it is also possible to simulate the value of a ship domain, when other ships reach this distance and receive a collision risk warning, the sailor must consider whether or not to take action immediately. There is also a collision risk between the two ships at this time, therefore, the sailor should be on the lookout and be prepared to take measures at any time to avoid a collision. Thus, the length of the major axis of the ship domain can determine the collision risk, which can provide an early warning to help sailors in the subjective decision-making process to avoid collisions.

In the past, most ship domain models have been too complex since many factors affect the ship domain. Researchers have previously chosen to ignore certain factors while retaining more important factors based on experiential knowledge. For most situations, own ship factors such as ship length, ship speed, and encounter angle have an important influence on the size of the ship domain owing to the limitations of the single simulation method. Environmental factors are often viewed as less important and abandoned due to limitations on the number of factors required. Nevertheless, environmental factors have some influence on the ship domain. Although wind and wave have little influence, factors such as fairway visibility and navigation density have a greater impact on the size of the ship domain. Therefore, the influence of certain environmental factors on the ship domain cannot be ignored. Taking these factors into account undoubtedly reduces experimental errors in determining the size of the ship domain, which makes the ship domain a more valuable warning signal. For example, when only the ship's own factors were considered in the present study, the ship domain size was D1 = 3190 m. In this case, the ship would receive the danger warning at 1595 m, instead of 1495 m, and take unnecessary measures to avoid collision. Premature preventive measures waste time, manpower, and material resources. Fuzzy reasoning can be used to synthesize various factors and produce more accurate simulation results, leading to fewer errors, which can save both time and unnecessary labor.

The traditional early warning methods of ship collision mainly rely on the ship's Collision Risk Index (CRI) to complete. It refers to the possibility of collisions between ships. Generally, when CRI = 0, it means that the ship is far away from the target ship and there is no danger of collision; when CRI = 1, it means that no matter what collision avoidance actions are taken, a collision will occur between the two ships. Most studies took CRI = 0.5 as the basis of whether to take action when establishing a ship collision avoidance decision model. If CRI > 0.5, then collision avoidance action is required. If CRI < 0.5, there is no danger of collision, just lookout.

The calculation of CRI usually depends on the distance of closest point approach (DCPA) and the time of closest point approach (TCPA). Many researchers have adopted different methods to calculate CRI by input DCPA and TCPA. Some of them adopted fuzzy inference methods [23] and some adopted fuzzy comprehensive evaluation [24]. The calculation of TCPA and DCPA is the first thing to do. The calculation process is shown as follows. Figure 13 is a diagram of ship relative motion parameters.

$$DCPA = R_t \cdot \sin(\varphi_\tau - a_t - \pi) \tag{5}$$

$$TCPA = R_t \cdot \sin(\varphi_\tau - a_t - \pi)/v_\tau, \tag{6}$$

where v_τ is the relative speed of the target ship, φ_τ is the angle of the relative speed of the target ship, R_t is distance between two ships, α_t is the true position of the target ship relative to own ship. The calculation formulas are as follows.

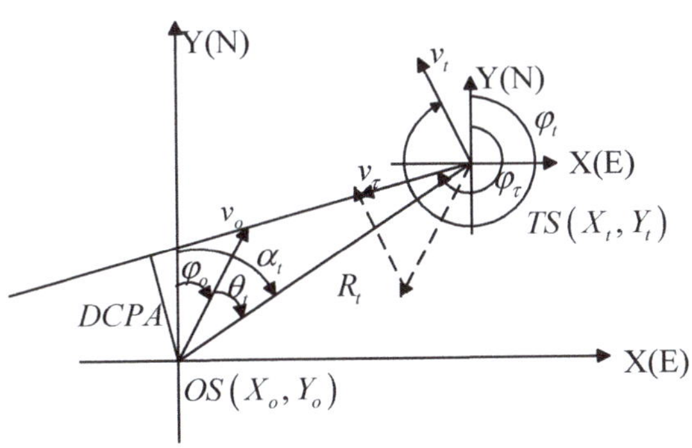

Figure 13. Diagram of ship relative motion parameters.

The specific solution process will not be introduced in detail, but the calculation of the relative speed and the relative course between the two ships requires the speed and course of the own ship and the target ship. This means that adopting this method to judge the collision risk index should obtain not only the speed and course of the own ship, but also that of the target ship.

As can be seen from the article and the simulation results, when the composition fuzzy inference based on the ship domain is adopted, only the ship's and environmental information at a certain moment need to be obtained, and there is no need to judge the status of the target ships, even static obstacles. As long as there are obstacles within the danger warning range during sailing, it will receive the collision warning and need to take avoidance actions. This makes the judgment of collision risk simpler and faster, and can also solve the limitation of judging only dynamic obstacles when judging the collision risk based on the CRI.

When judging the collision risk adopting the traditional method, the result obtained is the CRI of the target ship relative to own ship at a certain moment and position, and it can only be known whether it is necessary to take action at this time. However, the collision risk distance of the ship at a certain moment is obtained by the composition fuzzy inference. The method is more intuitive, more applicable, and has a certain predictability, which can give the sailor more reaction time.

At the same time, the data provided by the AIS system required by the traditional method does not contain environmental factors, and to a certain extent the impact of environmental factors on ship collisions is ignored. The composition fuzzy inference method has smaller errors.

The comparison table between composition fuzzy inference and traditional methods is shown in Table 4.

Table 4. The comparison table between composition fuzzy inference and traditional methods.

Early Warning Method of Ship Collision		Output	Input	Factors Considered		
				Own Ship's Factors	Target Ship's Factors	Environmental Factors
Composition fuzzy inference		Danger warning distance	Ship length ship speed encounter angle navigation density fairway visibility	√	×	√
Method for judging collision risk index	Fuzzy inference	CRI	DCPA TCPA	√	√	×
	Fuzzy comprehensive evaluation			√	√	×

5. Conclusions

In this paper, the ship domain was regarded as the safe domain for avoiding collisions between vessels. To establish the model, the ship domain was represented as a fixed ellipse and certain factors affecting the ship domain were considered. Not only the ship's own factors that have a greater impact on the ship domain are selected but also the environmental factors that have relatively small impact on the ship domain are added. Then, composition fuzzy inference of the selected factors was used to determine collision risk according to the range of the ship domain. The simulation results provide a reference for implementing collision avoidance actions in different ship encounter situations, which can improve the subjective decision-making of sailors and reduce the probability of collision. Compared with traditional methods of studying ship collision danger warning, the research process of collision danger of composition fuzzy inference is simpler and faster, the result is more intuitive, more applicable, and with fewer errors and has a certain predictability, which can remind the sailor to make preparations in advance.

Author Contributions: Conceptualization, Z.G. and H.W.; methodology, Z.G.; software, Z.G.; validation, Z.G., H.W., Y.W., and Z.Z.; formal analysis, Z.G.; investigation, Z.G.; resources, Z.G.; data curation, Y.W. and Z.Z.; writing—original draft preparation, Z.G.; writing—review and editing, Z.G.; visualization, Z.G.; supervision, H.W.; project administration, H.W.; funding acquisition, H.W. All authors have read and agreed to the published version of the manuscript.

Funding: This research received no external funding.

Conflicts of Interest: The authors declare no conflict of interest.

References

1. *Annual Overview of Marine Casualties and Incidents 2017*; European Maritime Safety Agency: Lisbon, Portugal, 2018; pp. 15–41.
2. Xiang, Z.; Hu, Q.Y. A calculation method based on AIS data for ship domain in restricted waters. *J. Transp. Eng.* **2015**, *15*, 110–117.
3. Hansen, M.G.; Jensen, T.K. Empirical Ship Domain based on AIS Data. *J. Navig.* **2013**, *66*, 931–940. [CrossRef]
4. Zhang, L.Y.; Meng, Q. Probabilistic ship domain with applications to ship collision risk assessment. *Ocean Eng.* **2019**, *186*, 106130. [CrossRef]
5. Ding, F. Research and Analysis of Open Water Ship Domain Based on AIS Data. Master's Thesis, Dalian Maritime University, Dalian, China, 2014.
6. Li, J.X.; Wang, H.B. Ship's Trajectory Planning Based on Improved Multiobjective Algorithm for Collision Avoidance. *J. Adv. Transp.* **2019**, *2019*, 4068783. [CrossRef]
7. Li, J.X.; Wang, H.B.; Guan, Z.Y. Distributed Multi-Objective Algorithm for Preventing Multi-Ship Collisions at Sea. *J. Navig.* **2020**, *73*, 971–990. [CrossRef]
8. Tsou, M.-C.; Hsueh, C.-K. The study of ship collision avoidance route planning by ant colony algorithm. *J. Mar. Sci. Technol.* **2010**, *18*, 746–756.
9. Lee, S.M.; Kwon, K.Y. A fuzzy logic for autonomous navigation of marine vehicle satisfying COLREG guidelines. *Int. J. Control Autom. Syst.* **2004**, *2*, 171–181.
10. Grinyak, V.M. Fuzzy collision avoidance system for ships. *J. Comput. Syst. Sci. Int.* **2016**, *55*, 249–259. [CrossRef]
11. Namgung, H. Inference System of Collision Risk using Ship Near-Collision Data. *J. Korean Inst. Intell. Syst.* **2019**, *29*, 395–402. [CrossRef]
12. Fujii, Y.; Tanaka, K. Traffic Capacity. *J. Navig.* **1971**, *24*, 543–552. [CrossRef]
13. Goodwin, E.M. A Statistical Study of Ship Domains. *J. Navig.* **1975**, *28*, 328–344. [CrossRef]
14. Davis, P.V.; Dove, M.J.; Stockel, C.T. A Computer Simulation of Marine Traffic Using Domains and Arenas. *J. Navig.* **1980**, *33*, 215–222. [CrossRef]
15. Coldwell, T.G. Marine Traffic Behaviour in Restricted Waters. *J. Navig.* **1983**, *36*, 430–444. [CrossRef]
16. Vander Tak, C.V.; Spaans, J.A. A model for calculating a maritime risk criterion number. *J. Navig.* **1977**, *30*, 287–295. [CrossRef]
17. Pietrzykowski, Z.; Uriasz, J. The Ship Domain in a Deep-Sea Area. In Proceedings of the 3rd International Conference on Computer and IT Applications in the Maritime Industries, Siguenza, Spain, 9–12 May 2004; Elsevier Science: Amsterdam, The Netherlands, 2004; pp. 204–211.
18. Pietrzykowski, Z. Ship's Fuzzy Domain -A Criterion for Navigational Safety in Narrow Fairways. *J. Navig.* **2008**, *61*, 499–514. [CrossRef]
19. Pietrzykowski, Z.; Uriasz, J. The Ship Domain -A Criterion of Navigational Safety Assessment in an Open Sea Area. *J. Navig.* **2009**, *62*, 93–108. [CrossRef]
20. Qu, X.; Meng, Q.; Li, S. Ship Collision Risk Assessment for the Singapore Strait. *Accid. Anal. Prev.* **2011**, *43*, 2030–2036. [CrossRef] [PubMed]
21. Wang, N.; Tan, Y.; Liu, S.M. Ship Domain Identification Using Fast and Accurate Online Self-organizing Parsimonious Fuzzy Neural Networks. In Proceedings of the Chinese Control Conference, Yantai, China, 22–24 July 2011.
22. Zhou, D.; Zheng, Z.Y. Dynamic Fuzzy Ship Domain Considering the Factors of Own Ship and Other Ships. *J. Navig.* **2019**, *72*, 467–482. [CrossRef]
23. Yao, J. Fuzzy decision of ship collision risk. *J. Dalian Fish. Univ.* **1998**, *13*, 29–34.
24. Wang, Q.F. Fuzzy decision model of collision risk and realization. In *Collection of Excellent Papers on Chinese Navigational Science and Technology*; Shanghai Pujiang Education Press: Shanghai, China, 2013; pp. 172–177.

MDPI
St. Alban-Anlage 66
4052 Basel
Switzerland
Tel. +41 61 683 77 34
Fax +41 61 302 89 18
www.mdpi.com

Journal of Marine Science and Engineering Editorial Office
E-mail: jmse@mdpi.com
www.mdpi.com/journal/jmse

www.ingramcontent.com/pod-product-compliance
Lightning Source LLC
Chambersburg PA
CBHW061603210926
51850CB00009B/2202
9783036559193